The Energy Transition and Climate Change

by
Kurt Olzog

The Energy Transition and Climate Change

Developments and Future Perspectives

Second Edition

Author: Kurt Olzog

During the course of his studies in mathematics and geography for his postgraduate teaching qualification, among other things the author engaged intensively with the development of the energy economy and wrote the first examination paper in this area.

In his years as a senior teacher at private schools and as a lecturer in the private sector, as well as management tasks and freelance organisational and business consultation, he followed the development of the energy economy in the daily and weekly press and public media as well as in the relevant literature.

Over the course of decades, signs of the first consequences of the intensive use of fossil fuels (coal, crude oil and natural gas) began to appear: Effects on the climate became apparent.

After the time-intensive consulting projects came to a close, the author was again able to focus more intensively on the energy economy and clearly present developments in energy production using fossil fuels, including uranium, along with increasingly important renewable energies.

In comparison to this, climate development over the last one hundred years and the dependency of the climate on the type of energy consumption by humans is presented quite vividly.

Bibliographical information of the German National Library:

The German National Library lists this publication in the German National Biography; detailed bibliographical information is available on the Internet at www.dnb.de.

TWENTYSIX - The Self-publishing Publisher

A cooperation between Random House publishing group and BoD - Books on Demand

© 2016 Kurt Olzog

Produced and published by:

BoD – Books on Demand, Norderstedt

The German edition was published with:

ISBN: 9783740710057

The English second edition is published with:

ISBN: 9783740730604

Contents

1. Development of the Energy Economy   pg. 6
2. Development of the Nuclear Economy   pg. 28
3. Development of Renewable Energy Sources   pg. 43
4. Climate Development over the Last Century   pg. 52
5. Future Perspectives   pg. 67
6. Paris Climate Conference 2015   pg. 100
Bibliography   pg. 111

1. Development of the Energy Economy

Fossil energy sources have increasingly been used for heating, electricity and locomotion since the industrial revolution. Oil in particular has become the most important source of energy for the world economy over the last century. Its share of world energy consumption in 1976 was almost 45 %, whereas all solid fuels (coal and lignite, peat, etc.) only accounted for 30% and natural gas not even 18%.[1]

As oil began to be in demand for industry in the second half of the 19th Century (in the United States and Russia, the petroleum industry developed at almost the same time), the demand for this versatile and inexpensive raw material grew at an ever increasing rate. Especially in North America, oil was consumed with increasing intensity, so that the rapidly increasing demand for oil led to an expanding oil industry. In particular, the car boom after 1911, due to which the car became a means of transport for the ordinary man, provided the oil companies with a continually expanding market, so that in the 1920s and 1930s, the search for oil began to expand throughout the entire world.

---

[1]   The British Petroleum Company Ltd., 1976, pg. 16

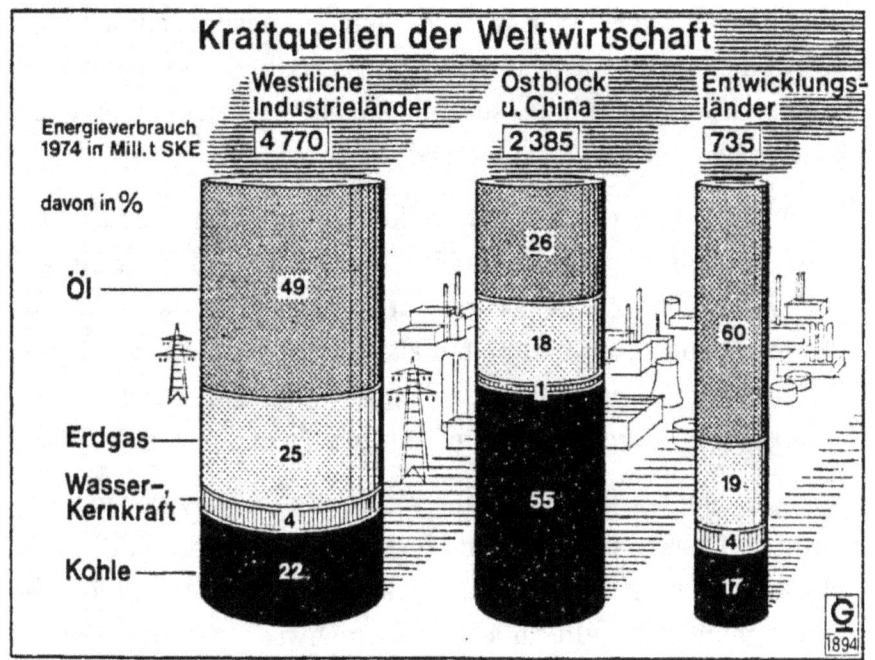

Taken from: Evers 1976, p. 106

In Iran and Iraq, in Venezuela and in Indonesia, oil was soon being extracted and exploration was intensifying at an increasing rate.

The USA, however, continued to be known as the land of oil between the two world wars, because on the one hand it had large oil reserves and on the other it also had a strong oil industry due to its extensive oil consumption. Shortly before the outbreak of the Second World War, Kuwait and Saudi Arabia were also able to begin exploiting the huge deposits discovered there.

The Second World War interrupted the promising activities of the oil companies in the Middle East. Instead, the American oil reserves were used up to such an extent that American oil exports had to be gradually discontinued.

After the war had ended, oil production in the Middle East gained new impetus, particularly as North America was increasingly becoming a deficit area. Thus, not only the sharp rise in Western European oil consumption, but also the increasing importation of oil to what was at the time the largest producing country, the USA, had to be covered by oil from Venezuela and the Middle East.

In quick succession, huge oil reserves were discovered in the Middle East, so that by the mid-1950s the proportion of Middle Eastern oil reserves in relation to all of the oil reserves discovered throughout the world came to more than sixty percent.

In the USA and the United Kingdom, flourishing oil companies were developing increasingly high-handed methods in their activities, which in turn contributed to the Iran crisis (1951 - 1954). The unsuccessful attempts to emancipate Iran may initially have intimidated the other oil-producing countries, but increasing Soviet influences in the Arab region offset the power of the industrial countries and the multinational oil companies working for them (one only has to think of Egypt in the 1950s).

The Suez crisis caused by Gamal Abdel Nasser in 1956 bears testament to the gradually shift in power; the former colonial powers of England and France in the Middle East and North Africa were obviously losing importance. In the meanwhile, the oil-producing countries realised that joint advocacy of their interests made them less defenceless against the arbitrariness of the industrial countries and their oil companies than isolated attempts at resistance.

Thus, in 1960, OPEC (Organization of the Petroleum Exporting Countries) was formed. The oil-producing countries initially used this new instrument to enforce more stable renvenues against the oil companies for their organization. Later, immediately after the Six Day War with Israel in 1967, they tested their first oil embargo against the United States, the United Kingdom and the Federal Republic of Germany.

But despite the embargo lasting for three months, it had little effect, firstly because of policies in the countries concerned, which at that time provided for the stockpiling of reserves, meaning that the embargo could be bridged for a certain amount of time, and secondly because of the additional demand for Venezuelan and Iranian production, which increased by a factor of multiples.[2] The result was that this event was seen more as a peripheral phenomenon to the Middle East war staged by the powerless Arabs.

In particular, this led to the policy of stockpiling being abandoned, as it was assumed that the oil-producing countries would no longer use the embargo method because of its ineffectiveness and the disadvantages it caused for the oil-producing countries themselves.

---

2   Lieser, 1975, p. 30 f

At the beginning of the 1970s, OPEC suddenly began to attract attention: oil prices were rising. This was repeated on a regular basis, which provoked a wave of indignation in the public spheres of the western industrial countries each time. This development reached its climax after the outbreak of the Fourth Middle East war, the Yom Kippur War on the Jewish holiday of Yom Kippur, 06 October 1973, during which Egypt took back a large part of the territories it lost in Sinai during the Six Day War, including major oil fields.[3]

Even when the former Soviet Union discovered extensive oil deposits in Western Siberia at the beginning of the 1970s, the Middle Eastern share did not fall below 50 percent and then increased again slightly after that. To this day, the Middle East is the most important oil-producing area, which is also reflected in the extent of production.

The weapon of the oil embargo was applied again and caused major panic among the oil-importing countries, especially in the USA, Japan and Western Europe, where oil consumption had increased from 1.5 billion tonnes in 1967 to more than 2.3 billion tonnes in 1973.[4] The oil-producing countries went one further: In order to curb oil production by 12 %, at conferences held in quick succession during a three month period, they agreed to gradually increase the price of oil by 400 percent.[5]

This triggered such an immense public shock in the western

---

[3] Oktoberkrieg und Truppenentflechtung (October War and Disengagement of Troops), Spiegel No. 32, 1978, p. 201
[4] The British Petroleum Company Ltd. (BP), 1976, p. 20
[5] Lieser, 1975, p. 21

industrialised countries that the resulting political and economic turmoil caused economic growth to falter in these industrialised countries: "Trade and balance of payments instruments are falling into disarray, inflation rates are increasing, growing unemployment is rife, the gross national products of the western industrialised countries are showing only minimal growth and each time the threat of war in the Middle East becomes real again, the lachrymose voices of politicians, the public media and the citizens concerned as sensitive energy consumers are raised."[6]

While peace efforts were in progress in the Middle East, the western industrialised countries were making increasing efforts to analyse the oil crisis or, as it was increasingly being called, the energy crisis and its causes and consequences, in order to better be able to face similar developments in the future. Thus, the International Energy Agency (IEA), a sub-organisation of the OECD, was set up.

This IEA was to be an instrument for the industrialised states in order to both secure against inconveniences on the part of OPEC and to seek and continue to develop a dialogue with the oil-producing countries, and thirdly to apply alternative energy sources in a more targeted manner than before on order to achieve greater independence from OPEC.

The oil crisis achieved one more thing: The attention paid to developing countries poor in raw materials intensified. Due to the huge increase in oil prices, these poorest of developing countries in particular experienced payment difficulties, with the result that

---

6 Ibid.

their loans skyrocketed, leaving them barely able to even pay the interest at times.

Thus, OPEC and IEA tried with all means available to assist these already vulnerable countries, which has been seriously damaged by the negative development of the global economy and by their own apathy (as well as their mass malnutrition).

The energy-economic interests of the most important members of the IEA have always varied considerably. Japan had to import all of the oil it consumed, which accounted for more than 74% of its annual energy consumption in 1974.

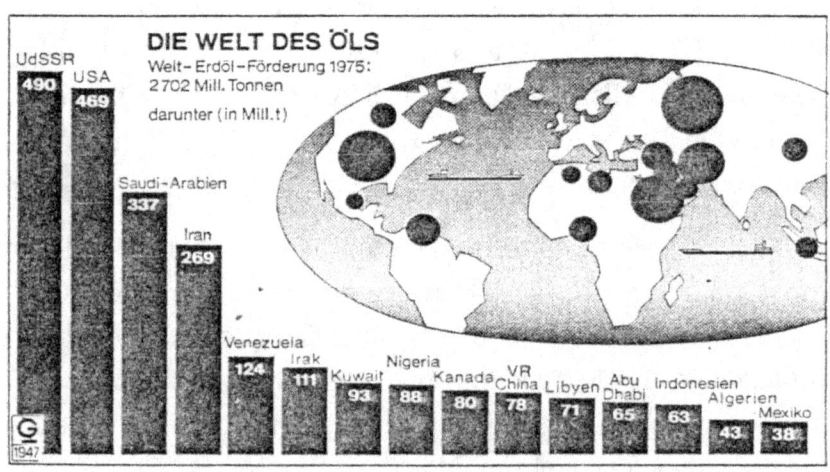

Taken from Fernau 1976, p. 94

The Japanese public had serious reservations regarding the development of nuclear energy at that time. The American economy, which was traditionally only slightly dependant on imports, was and is not as dependant on foreign markets as the Western European or Japanese economies. In the general political arena, however, the United States, as a leading power in the western world, is particularly sensitive to interactions between economic problems and its freedom to act in regard to foreign policy. Therefore, not only economic, but even more so general political interests motivate the United States to cooperate within the Energy Agency.

The Western European attitude to the IEA was quite differentiated. From the end of the 1960s, considerable reserves of oil were discovered in the British and Norwegian North Sea. In addition to this, the United Kingdom and the Federal Republic of Germany have considerable coal deposits. Moreover, the Netherlands and to some extent the United Kingdom have natural gas deposits, which cover almost half of the Netherland's primary energy consumption. In contrast to this, France and Italy are very poor in their own resources from primary energy sources and have to cover a significant share of their primary energy consumption through imports.

The role of the oil and gas companies has changed considerably since the oil crisis. "The oil companies no longer own the crude oil extracted in the OPEC countries. On the one hand, they became crude oil buyers, whereby they were able to contractually secure some long term purchasing options; on the other hand, they became service providers - again on a contractual basis - conducting crude oil production and exploration activities for the oil-producing countries. "[7]The fact that the oil companies passed OPEC's price dictates on to the buyers was simply a necessity.

The tasks involved in exploring and exploiting new oil fields, which were also becoming more and more expensive and required ever more complex technologies, demanded huge sums of investment capital, so that the concerns only acted with greater determination during the crisis, even if they did contribute a little to the price increases. This was not the first time alternative energy sources were discussed, and not just nuclear energy, which had at times caused unease even before the oil crisis. The expected exhaustion of oil reserves (at a constant production rate of three billion tonnes per year for the oil discovered up to this point, these would last for another 30 years),[8] forced alternative methods for the production of energy to be reconsidered.

In particular, the more efficient use of energy was to be ensured. The extensive wasting of energy in North America, which has always been the norm, no longer presented a role model for Western European energy policy. However, the efficiency of energy use was also too low here. This efficiency was therefore to be increased.

With the help of alternative energy sources and more efficient energy use, the world economy managed to reduce oil consumption over the course of the period from 1975 to 1985. The decrease in production was linked to the sustained restructuring of world oil production. In Western Europe, North America and

---

7   Burchard 1076, p. 124
8   BP 1976, p. 4, 20

Africa, production increased, while in the Middle East production was greatly reduced, mainly to keep the price of oil stable. Here, production decreased from almost 970 million tonnes to around 506 million tonnes.[9]

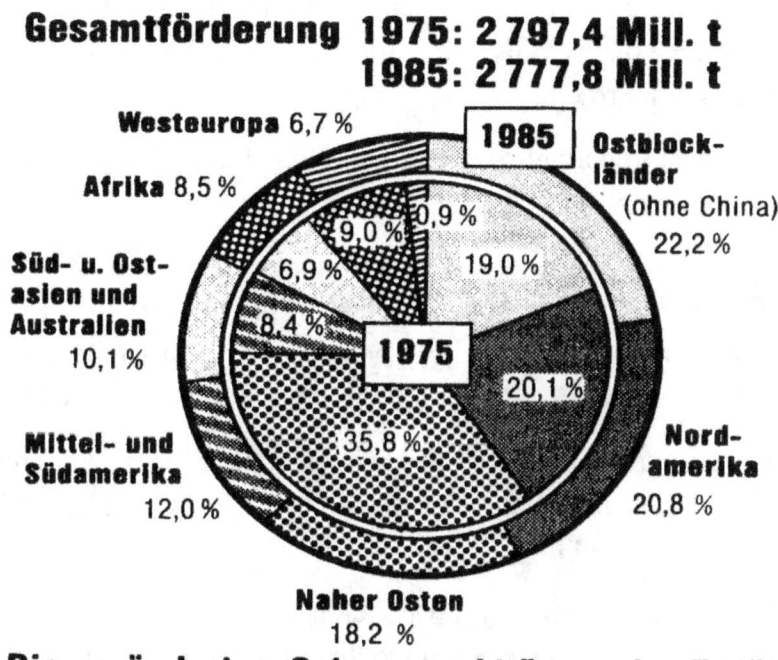

Die veränderten Schwerpunkträume der Erdölförderung 1975–1985

"The share of OPEC countries... in world oil production came to just 29% in 1985 (as opposed to 54% in 1973). This corresponded to less than half of the production capacity of these countries, which was set to stabilise the price by preventing further

---

9   Der Fischer Weltalmanach 1987, p. 861f, with figure

oversupply."[10] The oil market had become a buyer's market with surpluses of supply."

"Worldwide consumption of primary energy rose by 34.5% from 1970-1980 and by 21.7% from 1980-1990. This slowdown in the increase in consumption continued into the 1990s; consumption was then increasing on an annual basis by about 1 % and was therefore considerably below global economic growth and global population growth.

The most important source of energy by far in global terms is oil (1993: 36.8 %).[11]

The increase in the consumption of natural gas since the 1970s was significantly stronger than the overall increase in energy consumption. Its share increased from 19.5% in 1970 to 24.0% in 1993.

The proportion of fixed energy raw materials, coal and lignite (the second most important energy sources), fell from 32.9% in 1970 to 28.9% in 1993.

"The highest growth rates were demonstrated by the use of nuclear energy, especially up to the mid-1980s (1970: 0.1 %; 1980: 1.2 %; 1985: 5.5 %; 1990: 6.8 %). In the last few years, its share has stagnated at 7.2%."[12]

---

10   Ibid. p. 863
11   Der Fischer Weltalmanach 1997, p. 1052
12   Ibid. p. 1053f with table

Use of energy sources for global energy consumption 1970-1992
(only commercial energy), see "Yearbook of World Energy Statistics", UNO, billions of tce (coal units, Mrd. t SKE, Steinkohleeinheiten)

	1970		1980		1990		1992		1993	
	Mrd. t SKE	%	Mrd. t SKE	%	Mrd. t SKE	%	Mrd. t SKE	%	Mrd. t SKE	%
Erdöl	3,009	45,3	3,835	44,6	4,011	36,9	4,028	36,7	4,074	36,8
Kohle	2,184	32,9	2,623	30,5	3,239	29,8	3,226	29,4	3,207	28,9
Erdgas und Stadtgas	1,293	19,5	1,836	21,4	2,563	23,6	2,596	23,7	2,659	24,0
Kernenergie	0,010	0,1	0,101	1,2	0,738	6,8	0,792	7,2	0,806	7,2
Wasserkraft, Sonstige	0,145	2,2	0,198	2,3	0,314	2,9	0,319	3,0	0,339	3,1
Insgesamt	6,641	100,0	8,593	100,0	10,865	100,0	10,961	100,0	11,085	100,0

The Table shows the most important sources of energies and the development of their consumption. The calculations of coal units are difficult and can differ in different sources, dependant on the mode of calculation and the amount of non-commercial sources of energies as animals for transportation, firewood etc.

Over the next ten years, consumption again increased by about 11%, whereby the importance of oil decreased by 34.3% in 2002 due to the growth in natural gas. Natural gas and town gas had drawn equal with coal (hard coal and lignite) in terms of importance, accounting for slightly more than 27% of total consumption.[13]

Use of energy sources for global energy consumption 1970-1992
(only commercial energy)

	1970		1980		1990		2000		2002	
	Mrd. t SKE	%	Mrd. t SKE	%	Mrd. t SKE	%	Mrd. t SKE	%	Mrd. t SKE	%
Erdöl	3,009	45,3	3,835	44,6	4,011	36,9	4,311	35,3	4,361	34,3
Kohle (Stein- und Braunkohle)	2,184	32,9	2,623	30,5	3,239	29,8	3,217	26,4	3,496	27,5
Erdgas und Stadtgas	1,293	19,5	1,836	21,4	2,563	23,6	3,319	27,2	3,459	27,2
Kernenergie	0,010	0,1	0,101	1,2	0,738	6,8	0,947	7,8	0,981	7,8
Wasserkraft, Windkraft, Sonstige	0,145	2,2	0,198	2,3	0,314	2,9	0,404	3,3	0,401	3,3
Verbrauch Insgesamt	6,641	100,0	8,593	100,0	10,865	100,0	12,198	100,0	12,698	100,0

Source: Yearbook of World Energy Statistics, UN

---

[13] Der Fischer Weltalmanach 2007, p. 672f with table

Total consumption was also precipitated by the rapidly growing Chinese consumption, because in the meantime, China had become the second largest consumer of energy after the United States.[14]

The largest consumers of energy
in billions of tce

	2002	2001	2000	1990
USA	3177,8	3117,9	3167,2	2 686,9
VR China	1271,1	1096,4	1009,1	893,4
Russland	856,9	860,2	851,4	–
Japan	677,6	674,3	672,1	564,2
Indien	473,6	456,1	455,1	269,2
Deutschland	457,9	468,7	455,8	501,3
Kanada	352,8	348,3	354,2	291,9
Frankreich	348,8	345,9	333,2	294,7
Großbritannien	318,1	329,3	331,9	307,4
Italien	252,8	250,7	247,5	223,7
Rep. Korea	238,4	227,6	221,7	119,1
Südafrika	198,9	190,3	190,8	115,2
Mexiko	196,4	195,1	196,7	157,8
Ukraine	190,9	208,1	204,2	–
Brasilien	181,6	180,5	177,4	116,9
Spanien	165,9	158,4	156,5	80,8
Australien	160,9	166,3	157,3	127,1
Polen	120,3	123,9	122,5	96,0
u. a. Österreich	37,3	38,0	35,8	31,9
Schweiz	34,0	35	32,9	31,9

Source: UN

---

14 Ibid. p. 673 with table

India was also consuming more energy than Germany, for example.

During this period, there were Iraqi-Kuwaiti disputes over oil production in the shared border area. On 02/02/1990, Iraq occupied Kuwait and declared Kuwait the 19th Iraqi province. Following sanctions and UN resolutions, this led to the second Gulf War on 17/01/1991,[15] which resulted in Iraq's withdrawal but left burning oil fields behind. However, the world energy economy was barely shaken by this.

Ten years later, on 11/09/2001, the two high-rise buildings of the World Trade Center in New York and the Pentagon were destroyed or damaged by terrorist attacks. Iraq refused to condemn this terrorist attacks. This led to a further Iraq War on 20/03/2003 due to the intervention of the United States and the United Kingdom, this time without a UN mandate.[16] But this had little effect to the world energy economy.

Renewable energy sources such as hydro-power, solar and wind energy and biomass still did not play a significant role at this point in time, although it had been shown in the meantime that carbon dioxide, as a greenhouse gas, was the main contributor to global warming in the light of the quantities emitted.

Hydrogen as an energy source was only available to a limited extent and was generated using fossil energy sources rather than renewable energy sources such as excess solar or wind energy for

---

15   DIE ZEIT: Das Lexikon in 20 Bänden (The Lexicon in 20 Volumes), Hamburg 2005, Volume 07, p. 134
16   Ibid. p. 135

example. The first tentative steps were only taken a decade later. An attempt to convince the sun-drenched Arab countries that they could invest in preparing for the post-oil era resulted in the hilarious suggestion that they should be granted aid because the previously valuable oil would gradually lose value as it would no longer be used. The DESERTEC Concept, which was launched by well-known companies for this purpose, lost much of its persuasive power because of this.[17]

As early as 1972, a report by the Club of Rome[18] became known, under the title "The Limits to Growth", published by D. Meadows et al,[19] in which the principle of limited resources on our planet is convincingly described.

In the meantime, the prices of fossil fuels have increased so much that it has become worthwhile to exploit tar sands and shales found in North America.

---

17 More on this later
18 Club of Rome, initiated in 1968 by Aurelio Peccei (1908 to 1984), an informal association of scientists, politicians and business leaders from numerous countries
19 D. Meadows et al., 1972

Oil production, millions of tons

	2008	2012	2013
Saudi-Arabien	509,9	549,8	542,3
Russland	493,7	526,2	531,4
USA	302,3	394,1	446,2
VR China	190,4	207,5	208,1
Kanada	152,9	182,6	193,0
Iran	214,5	177,1	166,1
Ver. Arab. Emirate	141,4	154,7	165,7
Irak	119,3	152,5	153,2
Kuwait	136,1	153,7	151,3
Mexiko	156,9	143,9	141,8
Venezuela	165,6	136,6	135,1
Nigeria	102,8	116,2	111,3
Brasilien	98,8	112,2	109,9
Angola	93,1	86,9	87,4
Katar	65,0	83,3	84,2
OPEC	1 746,0	1 776,3	1 740,1
Weltförderung	3 993,2	4 119,8	4 132,9

Source: BP 2014

Oil and gas deposits exploited on the deeper seabed have also become worth exploiting. The newly applied chemical fracking procedures for gas and oil production from greater depths has even allowed the USA to become considerably less dependent on gas imports.[20]

---

20 Springer, Michael: Wird Fracking den Energiehunger stillen? (Will fracking quench the thirst for energy?) In: Spektrum der Wissenschaft 8/14 p. 20

"The proven and recoverable reserves of crude oil remained constant in 2013 in comparison to the previous year and, according to BP, came to 1687.9 billion barrels. The static range of crude oil is thus 53.3 years; in Europe (including Russia and the CIS states) 23.4 years, in the Middle East on the other hand, the range is 78.1 years. 72% of the reserves are accounted for by OPEC countries, around three quarters of which by the Middle East. The OECD countries on the other hand only account for 15%.

These figures underline the importance of OPEC and the Gulf region in particular for the future supply of crude oil. As in the previous year, in 2013 Venezuela was the most oil-rich country in the world with 18% of all confirmed reserves. Saudi Arabia accounted for 16%, Canada 10%, Iran and Iraq 9% each and Kuwait 6%. The reserves of non-conventional oil, such heavy oils in Venezuela, oil sands in Canada and Russia and oil shales in the United States and Canada, will play an increasingly important role in the energy supply of the future. Crude oil prices increased in the period from 2002-2008 to an extent not previously believed possible. The maximum price of crude oil was reached on 11/07/2008 at USD 147.50/barrel for North Sea grade Brent. This made oil five times more expensive than it was in 2002. The onset of the global economic and financial crisis in the middle of 2008 caused oil prices to fall to around USD 38 per barrel in December 2008. Since the beginning of 2009, the prices have clearly recovered and, at the beginning of 2011, exceeded the mark of USD 100 per barrel, around which they have remained ever since. The annual average prices for crude oil grade Brent for 2011 and

2012 are listed at around USD 111/barrel and showed a slightly decreasing tendency in 2013 to USD 108.66/barrel"[21].

Source: finanzen.net 2014

The problem of price regarding fossil fuels resulted in increased efforts in the development of nuclear energy. Uranium does not have to be highly enriched for peaceful use at nuclear power plants (up to 4% fissionable U235 in a mixture with U238). Normally, extracted uranium is 99.3% U238. In gas centrifuges, the proportion of U238 is reduced, so that the uranium mixture can be used for the desired chain reaction in a reactor. In the reactor, several U238 atoms catch slow neutrons, so that a certain proportion of plutonium Pu239 results, which in turn can be used for the chain reaction.[22]

---

21   Der neue Fischer Weltalmanach 2015, p. 662, with table p. 21 and graphic
22   Lexikon der Physik (Lexicon of Physics), 2000. Spektrum Akademischer Verlag GmbH Heidelberg, Volume 5 p. 348f, Volume 4 p. 294f

The possible "breeding" of Pu239 from U238 in so-called breeder reactors was one of the reasons to believe that the nuclear fuel uranium could still be used for many centuries to come. In the German translation released in 2003 of the "Global 2000 – Der Bericht an den Präsidenten (Global 2000 - The Report to the President)"[23], on page 72 et seq., the growth rate for nuclear energy development was forecast at more than 200 % by 1990. But the world population gradually found out that this technology was not as secure as the energy industry claimed, due to numerous near misses and super-GSA (GSA - Greatest Supposed Accident); increased to super-GSA upon core meltdown in the reactor).

Now we are writing about 2015, and the aforementioned climate change not only brings measurable, but also tangible annual global warming. As explained in detail below, the unrestrained use of fossil fuels such as oil, coal and natural gas leads to a significant increase in carbon dioxide in the earth's atmosphere. This creates a greenhouse effect around the world that is causing the glaciers to melt, the ice sheet at the North Pole to recede, and to become thinner in parts of the South Pole, Greenland and Alaska. The result is a gradual increase in the ocean level, so that smaller Pacific islands already have to be evacuated. This allow states that can afford it to introduce more effective energy use and to promote the use of alternative energy sources such as photovoltaics and wind energy. Together with the use of the aforementioned fracking technology and the use of tar sands in North America, the resulting surplus of oil that is reflected in the

---

23 US Foreign Ministry et al.: The Global 2000 Report to the President, Washington 1980

price of oil.

The weekly newspaper "DIE ZEIT" published an article on 08 January 2015 on the Petrobras oil company in Brazil, describing this matter.[24] After that, the price of oil fell by half within six months.

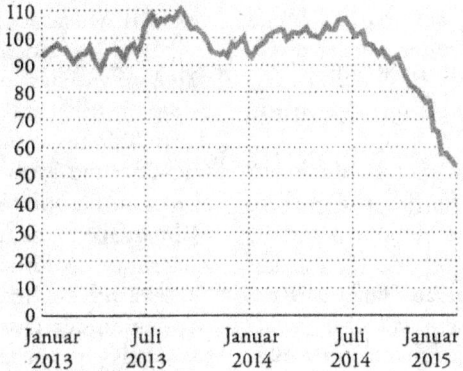

Graphic **ZEIT** / Source: Onvista

Because, on the one hand, the occurrence of fossil fuels is essentially limited and, on the other, the world climate is becoming disadvantageous to us through their use, the question arises as to whether energy production can be further developed based on a combination of renewable and climate-neutral energy

---

24 Thomas Fischermann: Es läuft wie schlecht geschmiert. (Its running like a poorly oiled machine) In: DIE ZEIT No 2 2015 p. 25, with graphic

sources such as photovoltaics and wind energy and increasing hydrogen economy (e.g. as a storage technology).

The fossil fuels could be stored in the ground and only extracted again in the event of more favourable environmental conditions. Lubricants and plastics can increasingly be manufactured using renewable raw materials.

At the moment, oil production is still increasing, while consumption is gradually stagnating. The price of oil is still low, so the motivation to change to renewable energy production is not easy to increase.

"The proven and recoverable reserves of crude oil remained constant in 2014 in comparison to the previous year and, according to BP, came to 1700.1 billion barrels. The static range of crude oil was thus 52.5 years; in Europe (including Russia and the CIS states) 24.7 years, in the Middle East on the other hand the range was 77.8 years. The revaluation of reserves in Venezuela in 2011 led to an increase in the static range of the oil reserves in Central and South America to over 100 years - the highest value worldwide."[25]

---

25   Der Fischer Weltalmanach 2016, p. 661f with table

Oil production, millions of tons

	2009	2013	2014
Saudi-Arabien	456,7	538,4	543,4
Russland	500,8	531,0	534,1
USA	322,3	448,5	519,9
VR China	189,5	210,0	211,4
Kanada	152,8	194,4	209,8
Iran	205,5	165,8	169,2
Ver. Arab. Emirate	126,2	165,7	167,3
Irak	119,9	153,2	160,3
Kuwait	121,2	151,5	150,8
Venezuela	155,7	137,9	139,5
EU	99,8	68,5	67,0
OPEC	1 622,6	1 734,4	1 729,6
Weltförderung	3 885,8	4 126,6	4 220,6

Source: BP 2015

2. Development of Nuclear Economy

Hiroshima: Ruins of the Chamber of Industry and Commerce building ('Atomic Bomb Dome'), Memorial for the Atomic Bomb Explosion in 1945

During the Second World War, nuclear fission was researched and developed to create the atomic bomb. At the end of the war in 1945, nuclear fission bombs were dropped on the Japanese cities of Hiroshima and Nagasaki. "The dropping of a US atomic bomb on Hiroshima on 06/08/1945 (the first ever use of nuclear weapons) resulted in about 200,000 deaths (many were the victims of late effects) and destroyed 80% of the city; reconstruction from 1949."[26]

After the war, the victorious powers developed and tested nuclear fusion bombs, also called hydrogen bombs, but fortunately these were not used. The use of nuclear energy for peaceful energy production since then up to the present day only referred to nuclear fission. The use of nuclear fusion for the production of energy was not yet commercially available. In the Federal Republic of Germany, the development of nuclear energy for peaceful energy production also began to be driven forward.

In 1956, a nuclear commission was formed to advise the federal government on legislative matters. This was later no longer required and was dissolved in 1971.[27]

"In Biblis am Rhein near Worms, the enormous structures not only surpass the venerable cathedral of the old imperial city in terms of their height and mass but also everything that has ever been built in the area.

---

26 DIE ZEIT: Das Lexikon in 20 Bänden (The Lexicon in 20 Volumes), Volume 6, p 423 with figure
27 Winnacker/Wirtz 1975, p. 76ff

They are the reactor domes and cooling towers of two nuclear power plants with a combined capacity of 2500 million watt. ...

These two large-scale power plants are the largest examples of the light water reactors generation to date. Such power plants are currently being built in many industrial countries and at various locations in West Germany. For the next one or two decades, they will be the first development level in the use of nuclear power."[28]

On 26/04/1986, the biggest nuclear reactor disaster in the history of civil nuclear power occurred in Chernobyl in the Ukrainian region of Chernobyl.[29]

"During a test on the turbo generators in Block 4, under changed reactor operating conditions, a sudden increase in performance could no longer be controlled and lead to several steam explosions and fires that completely destroyed the reactor."[30] A radioactive cloud formed from the leak in the damaged reactor core which extended as far as Scandinavia and Western Europe. Soviet information policy initially tried to play down the disaster. Help was only accepted from foreign specialists six days after the radioactive fall-out.

---

28   Ibid. p. 191
29   Der Fischer Weltalmanach 1987, pg. 216ff
30   DIE ZEIT: Das Lexikon in 20 Bänden (The Lexicon in 20 Volumes), Volume 15, p 110f with figure

Chernobyl: View of Block 4 of the nuclear power plant after the explosion on 26/04/1986

The then Soviet Secretary-General Gorbachev spoke of the incident for the first time in a televised speech on 14/05/1986. The combustion process in the graphite part of the reactor could only be stopped two weeks after the accident. Officially, 203 plant workers suffered serious radiation poisoning, 17 died from burns and 15 from radiation. 45,000 people were evacuated from the affected areas, followed by a further 90,000 inhabitants later on (official announcement).[31]

"The catastrophe occurred as an incorrectly designed test of the safety system was being carried out."[32]

The map on the next page shows the distribution of nuclear power plants in the former Soviet Union. The reactor at Chernobyl was a light-water-graphite reactor, also called a boiling water reactor (BWR). Reactors of this type were initially built mainly because of their simpler and cheaper construction. Then came pressurised water reactors, which had a more complex structure using separate water and steam circuits, but achieved a higher level of safety.

31　　Der Fischer Weltalmanach 1987, p. 217ff
32　　Ibid. p. 218f with map

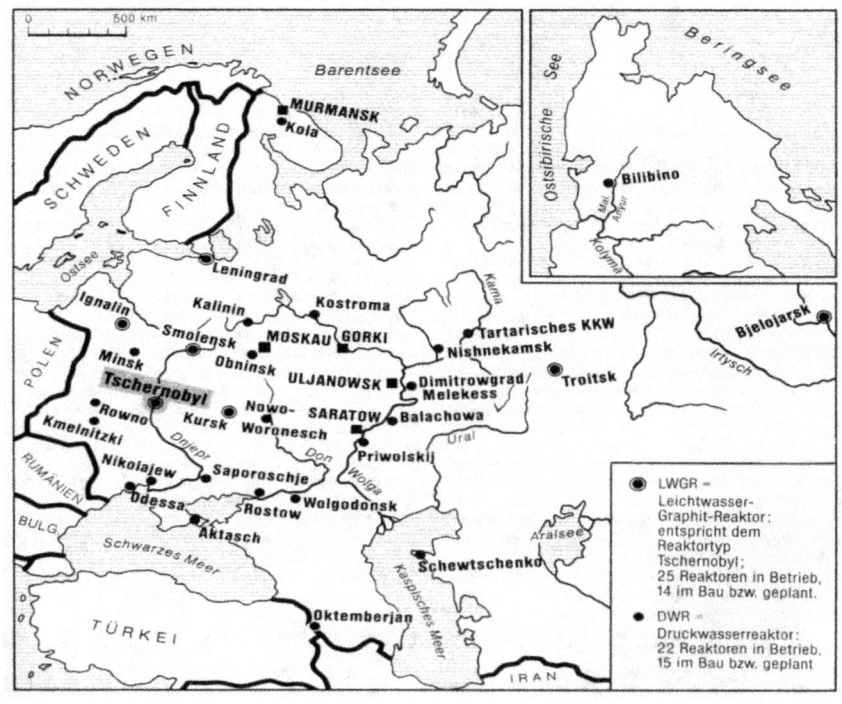

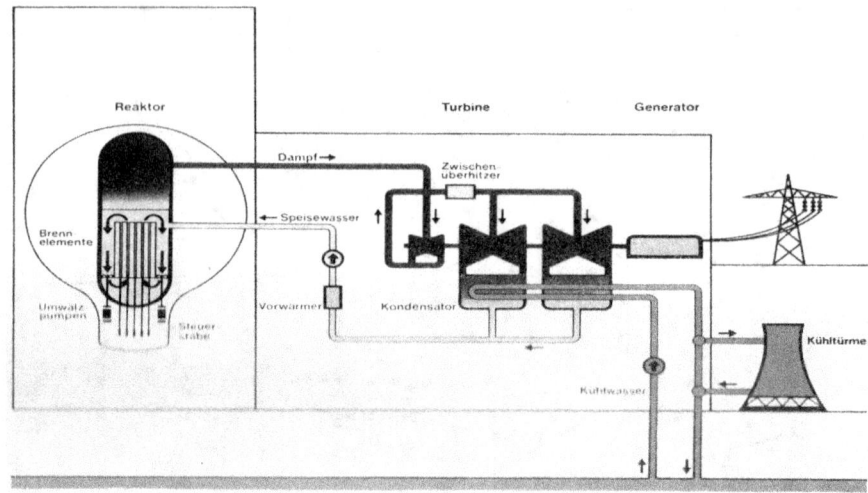

Circulations of a boiling water reactor (BWR) [5]

"The fundamental difference between the PWR and BWR is the fact that the BWR only has one circuit. The steam is generated directly in the reactor pressure vessel and sent straight to the turbine... In doing so, the reactor core remains covered with water. In the pressure vessel-mounted pumps, the water circulates over the core within the short circuit. Some of this water is evaporated in the process and replaced by feeding in condensed water...The advantages of BWR over PWR include a more simple structure, lower pressure in the reactor pressure vessel and somewhat greater efficiency. These advantages must, however, is offset against the radioactive contamination of the turbine, when is unavoidable due to the use of the steam produced in the reactor and makes work on the machine more difficult."[33]

---

33 Münch, Erwin 1980, p. 34f, with figure

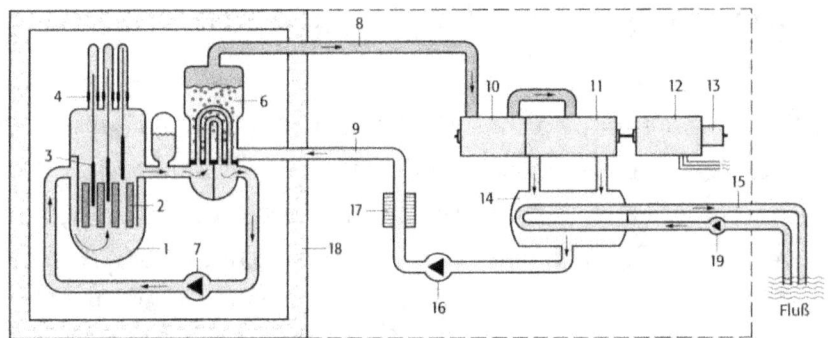

Pressurised water reactor: Schematic structure: 1) Reactor pressure vessel, 2) Uranium fuel elements, 3) control rods, 4) control rod drives, 5) pressuriser, 6) steam generator, 7) coolant pump, 8) live steam, 9) feed water, 10) high-pressure part of the turbine, 11) low pressure part of the turbine, 12) generator, 13) excitation device, 14) condenser, 15) river water, 16) feed water, 17) preheating system, 18) concrete shield, 19) cooling water pump.[34]

This figure schematically illustrates the PWR. "The heart of the reactor, the core, is located in the interior of a solid steel pressure vessel with a wall thickness of 20 to 30 cm. It consists of densely packed thin fuel rods containing the fuel made of uranium oxide in a metal case. Between the fuel rods, the control rods made of neutron-absorbing material move. The control rods are moved by electro-mechanical drives that are mounted on the cover of the pressure vessel. The rods are introduced into the core by gravity. The heat generated by nuclear fission in fuel rods is taken up by the water that is pumped between the fuel rods. The water also serves as a moderator. The water, which is at a temperature of 323°C, is routed to the steam generator. Here, it flows through a large number of small tubes, whereby the heat passes through the pipe wall to the water for the secondary circuit on the outside of the pipe. The primary water leaves the steam generator, cooled to

---

34   Lexikon der Physik (Lexicon of Physics), 2000, Volume 2, p. 97

approximately 290°C, and is sent back to the reactor pressure vessel.

The pressure in the primary circuit is so high at 155 bar that the water still does not evaporate despite being heated to 323 °C and is therefore called a pressurised water reactor. In contrast, the pressure on the secondary side is only at about 60 bar. At this pressure and the temperature the secondary water assumes in the steam generator, the water evaporates. The steam is used to drive the turbines."[35]

The super-GSA at Chernobyl had a significant impact on some neighbouring countries of the former Soviet Union: Poland, Hungary, Romania and Yugoslavia were openly critical of Soviet information policy. This led to protests demanding that the construction of nuclear power plants be abandoned and the building of nuclear power plants already decided upon be postponed. In the Federal Republic of Germany, the discussion about the use of nuclear power was reignited. In the Netherlands, plans for the construction of two new nuclear power stations were initially suspended.[36]

---

35   Münch, Erwin, 1980, p. 33
36   Der Fischer Weltalmanach 1987, p. 219

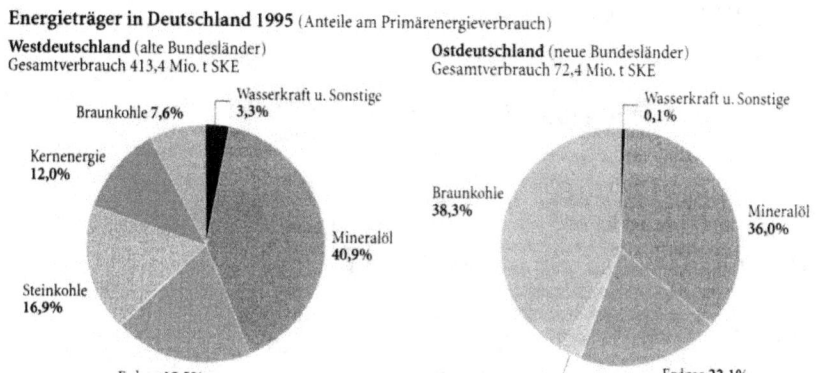

Source: Arbeitsgemeinschaft Energiebilanzen

Primary energy consumption in Germany in 1995 can be taken as an example of the further development of the use of nuclear energy. Here, the country refrained from constructing further nuclear power plants, so that nuclear power as an energy source in the old federal states only accounted for 12 % and was of hardly any significance at all in the new federal states. "In particular, the future contribution of nuclear energy (forecasts of between 8 - 15 % for electricity production) is uncertain, as political decisions play a greater role here."[37]

Another ten years on, the use of nuclear energy in comparison to primary energy consumption in Germany had hardly changed. In total, around 486 million tce (coal units) were consumed in 2005, the same amount as 1995. The share of nuclear energy increased slightly by 0.5% to 12.5% (see figure on the next page).[38]

---

37   Der Fischer Weltalmanach 1997, p. 1059-1062, with figure
38   Der Fischer Weltalmanach 2007, p. 675f, with figure

Energieträger in Deutschland 2005
(Anteile am Primärenergieverbrauch)

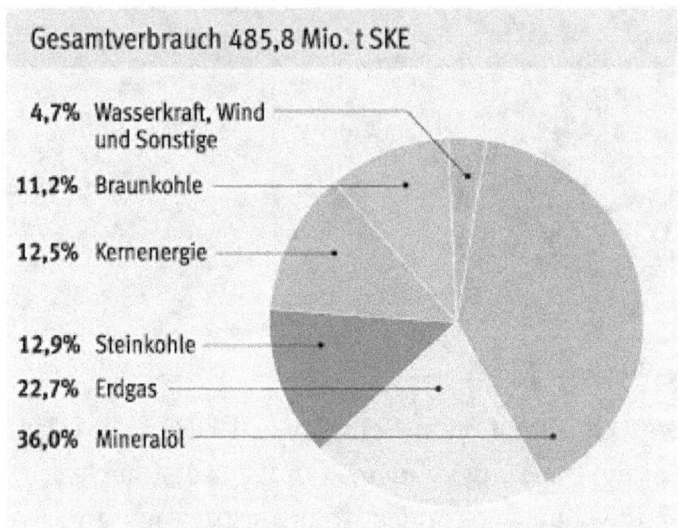

Source: Arbeitsgemeinschaft Energiebilanzen 2006

One reason for the stagnating development of nuclear energy in Germany was the SPD and "Alliance 90/Green Party" coalition, which was in government from 1998 to 2005 and during this time decided to phase out nuclear energy. The next big coalition between CDU/SPD did nothing to change this. As of 2009, the CDU/FDP coalition took over government and decided to reverse the decision to phase out nuclear energy. However, this decision did not last long: Until 2011.

"On the 11/03, an earthquake measuring 9.0 on the Richter scale - the worst in the history of Japan - shook the north-east of the country. The earthquake was followed by an enormous Tsunami that devastated large parts of the country...; according to the Japanese authorities, at least 15,000 people were killed and

500,000 were being housed at temporary shelters. As a result of the natural disaster, the cooling system at the Fukushima Daiichi power plant approximately 270 km north of Tokyo failed. Following several explosions, core meltdowns occurred in three reactor blocks. On 12/04, the Japanese Nuclear Safety Authority classified the disaster at Fukushima at the highest risk level 7 on the International Nuclear Event Scale (INES) - as high as the reactor disaster in Chernobyl in 1986."[39]

"The nuclear catastrophe at Fukushima... is the second event in the history of nuclear energy to be categorised as a **Level 7** incident".[40]

39	Der neue Fischer Weltalmanach 2012, p. 17, with figure
40	Ibid., p. 26, with figure on the next page

The destroyed reactor at the Fukushima Daiichi NPP on 24/03/2011

"The disaster at Fukushima also caused many countries to rethink atomic energy planning. Thus, by mid-2011, the following countries among others had announced a review of or **corrections of nuclear energy policy**:

• On 25/05/2011, the Swiss government decided to shut down the country's five reactors, which currently cover almost 40% of the electricity requirement, by 2034.

• In Germany, within the context of the **legislative package for the energy transition** adopted the Federal Parliament (30/06/2011) and the Federal Council (08/07/2011), it was decided not to recommission the seven reactors shut down during the course of the nuclear moratorium and NPP Krümmel, which had already been out of operation since 2009, and to shut down the remaining nine reactors at the latest by 31/12/2022.

• In Japan, on 10/05/2011, the government put its expansion plans for nuclear energy - from what is currently approximately 30% of the power supply to 50% by 2030 - on ice and announced an expansion of renewable energies.

• In Italy, the plans of the Berlusconi government for a return to nuclear energy were stopped by means of a referendum on 13/06/2011. 94.1% voted against nuclear energy, thus confirming the result of a referendum in 1987 following the Chernobyl disaster."[41]

The super-GSA in Fukushima had an effect worldwide:

"The **production of electricity by nuclear power stations** fell from 2,630 billion GWh (2010) to 2,518 billion GWh (2011), this corresponded to a decrease of 4.3%, the biggest since 1965. The main reason for this was the decrease in the production of electricity using nuclear energy in Japan (-44.3%) and Germany (-23.2%)".[42]

The following year, electricity produced using nuclear power around the world fell by 7%.[43]

"In 2013, the **production of electricity using nuclear power in Germany** decreased by -2.2% in regard to the preceding year according to AG Energiebilanzen. The nine nuclear power plants that remained on the grid generated 97.3 billion kWh (2012: 99.5 billion kWh), which corresponded to a 15.4% share in gross electricity generation in Germany (2012: 15.8%). This put nuclear

---

41   Ibid. p. 25
42   Der neue Fischer Weltalmanach 2013, p. 666
43   Der neue Fischer Weltalmanach 2014, p. 666

energy behind lignite, renewable energy sources and coal as the fourth most important energy source for the production of electricity. The share of nuclear energy in primary energy consumption was at 7.6% in 2013 (2012: 8.0%). The nine remaining nuclear power plants are to be decommissioned in the following order: Grafenrheinfeld (2015), Gundremmingen B (2017), Philippsburg 2 (2019), Grohnde, Gundremmingen C and Brokdorf (2021). The three newest installations, Isar 2, Emsland and Neckarwestheim 2, will be removed from the grid the latest by the end of 2022.

Whereas Germany, Switzerland and Belgium decided to phase out nuclear energy after the reactor disaster of Fukushima and nuclear energy was increasingly being critically discussed in Japan, the emerging countries in particular, such as China, Russia, India and Brazil, are pushing forward with the development of nuclear energy. Established nuclear nations such as the United States, Canada, the United Kingdom, Finland, Hungary, Slovenia, Slovakia and Sweden are keeping nuclear energy as part of their national energy mix and in some cases are also investing in new construction projects."[44]

On 23/07/2015, Deutschlandfunk, ZDF and ARD broadcast the news that France was planning to generate 50% of its power using nuclear energy instead of 80%. The rest is to be generated using renewable energy sources.

---

44   Der neue Fischer Weltalmanach 2015, p. 667

3. Development of Renewable Energy Sources

In the period from the end of the Second World War to 1985, renewable energy sources other than hydro-power played no role in global power generation. However, hydro-power plants were being built and further developed all over the world:

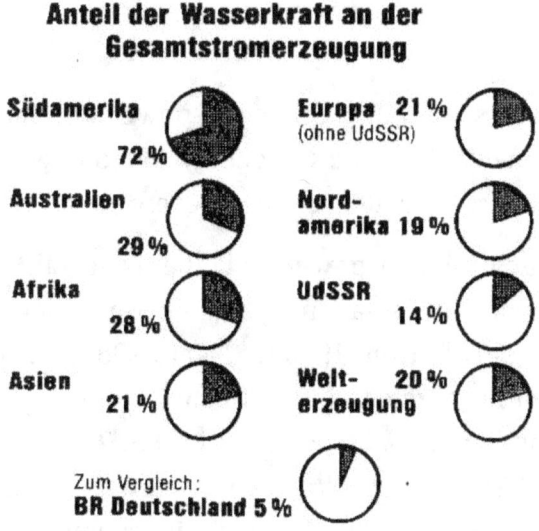

In the parts of the world with mainly developing countries, the production of electricity using hydro-power is significantly higher than in the industrialised countries.[45]

---

45 Der Fischer Weltalmanach 1987, p. 889, with figure

The development of the share of hydro-power in coal equivalents (tce) and as a percentage of global energy consumption from 1970 to 1983 are shown the following table:

Use of energy sources for global energy consumption 1970-1983
(see „ESSO" and „Yearbook of World Energy Statistics", UNO)

	1970		1980		1983	
	Mill. t SKE	%	Mill. t SKE	%	Mill. t SKE	%
Erdöl	3009	45,3	3990	45,6	3701	42,9
Kohle	2184	32,9	2625	30,0	2733	31,7
Erdgas	1293	19,5	1831	20,9	1855	21,5
Kernenergie	10	0,1	84	1,0	114	1,3
Wasserkraft	145	2,2	218	2,5	233	2,7
insgesamt	6641	100,0	8755	100,0	8635	100,0

Its share increased from 2.2 % in 1970 to 2.7 % in 1983.[46]

Twelve years later, in 1995, hydro-power still plays the main role among renewable sources of energy. Wind power and photovoltaics come under "other" energy sources.[47]

"The share of **hydro-power** and other **renewable energy sources** (such as wind and solar energy, geothermal energy) increased relatively slowly from 1970 (2.2%) to 1980 (2.3%) and then somewhat faster due to intensified development and targeted public funding in many countries (1985: 2.6% -1990: 2.9% -1995: 3.0% -2002: 3.3%). Hydropower is particularly important in many developing countries. In the industrial countries, apart from the mountainous countries of Austria, Switzerland and Norway, its share is relatively insignificant and difficult to increase. The use of other renewable energy sources such as solar and wind energy worldwide has not resulted in significant shares, even if they are

---

46   Ibid. p. 891f with table
47   Der Fischer Weltalmanach 1997, p. 1053f

of greater regional importance in some countries (e.g. geothermal energy in Iceland, wind power in Northern Germany and Denmark)."[48]

In Germany, renewable energy sources reached a share of approximately 4.7% of primary energy consumption in 2005 (see figure "Energy Sources in Germany 2005" in Chapter 2).

Three years later, the picture in regard to the generation of electricity according to energy source in Germany was significantly different:

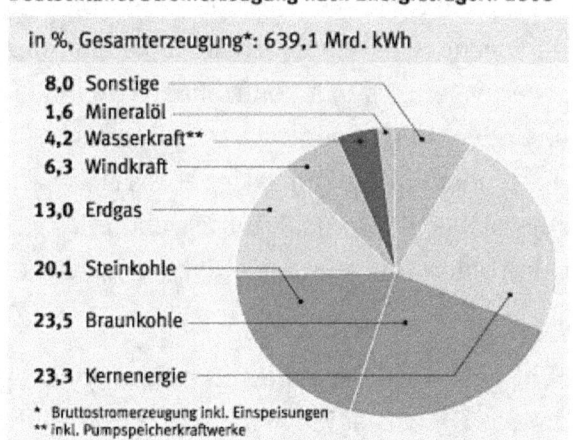

Source: Arbeitsgemeinschaft Energiebilanzen 2009

Together, hydro-power and wind energy had reached a share of about 10.5 % of electricity generation. Another 8 % was

---

48   Der Fischer Weltalmanach 2007, p. 673

accounted for by other sources, such as photovoltaics, biogas, etc.⁴⁹ Another two years later, in 2010, the share of renewable energy sources worldwide had again increased. "Renewable (or regenerative) energy sources - water, wind energy, biomass, photovoltaics, solar energy and geothermal energy - were becoming more important worldwide as complements to and substitutes for fossil fuels - oil, natural gas, coal and nuclear energy. More intensive use of renewable energy sources, with the exception of biomass, allows greenhouse gas emissions to be reduced, thus makes a contribution to climate protection. In addition to this, renewable sources of energy promote diversification of the raw materials base, reduce the dependence on fossil fuels, thus guaranteeing security of supply. Renewable energies are mainly domestic energy sources that contribute to regional value creation. In many developing countries, they can also provide large parts of the population with easier access to energy. To date, however, numerous technological, infrastructural, economic and political problems have made universal application difficult.

The worldwide **share of renewable energies in primary energy consumption** according to BP was at 7.8% in 2010. 6.5% of this was accounted for by hydro-power and 1.3% by the other renewable energy sources."⁵⁰

Taking power generation in Germany as an example, the development of the share of renewable energies is somewhat clearer:

---

49 Der Fischer Weltalmanach 2010, p. 703, with figure
50 Der neue Fischer Weltalmanach 2012, p. 682f

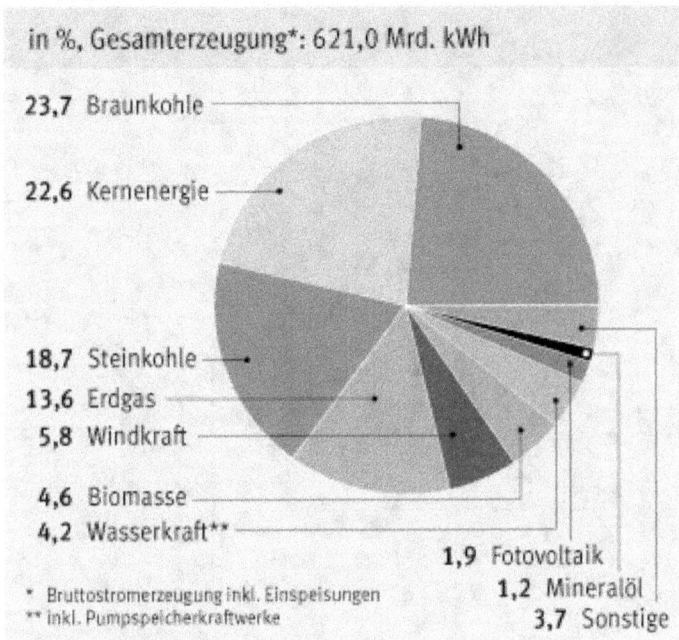

Source: AG Energiebilanzen 2011

Here, wind power, biomass, hydro-power and photovoltaics add up to a 16.5 % share.[51]

If we skip ahead three years, we again discover a rapid increase in the share of renewable energies supplying power in in Germany. Photovoltaics, at 4.7 %, supply more power than hydro-power (3.2%).

---

51 Ibid. p. 685 with figure

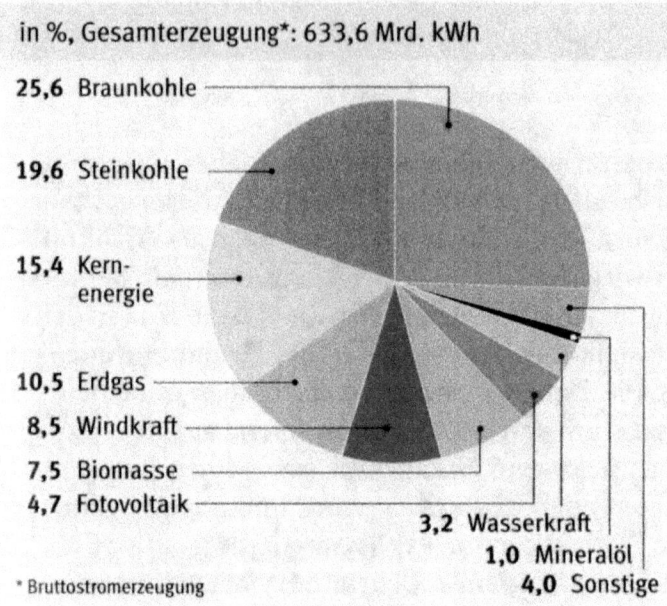

Source: AG Energiebilanzen 2014

Wind power, biomass, solar and hydro-power supply in a total share of 23.9% of electricity production.[52]

"Thus, renewable energies have become the second most important energy sources for electricity production after lignite."[53]

"Denmark achieved the highest share of renewably generated

---

52   Der Fischer Weltalmanach 2015, p. 669, with figure
53   Ibid. p. 668

electricity in 2013 at 47%, followed by Portugal (30%) and Spain (26%). If hydro-power is included in the calculations, the contribution of renewable energy sources to worldwide power generation in 2013 came to 21.7%. In Europe it accounted for 26.3%, in Central and South America 63.0% because of the high share of hydro-power; in the Middle East on the other hand, it only accounted for 2.6%."[54]

The share of renewable energy sources worldwide has increased significantly. In the five years from 2009 to 2014, it doubled to six percent. In Europe, including Russia and the CIS countries, the share in 2014 was as high as 10.5% and is still on the increase.[55]

This development was also reflected in Germany. "**Gross electricity generation** from solar, wind and hydro-power, biomass and household waste in Germany was at 160.6 billion kWh in 2014, +5.4% over the value for 2013. The contribution from renewable energy sources to gross electricity generation thus increased to 26.2% (2013: 24.1%), whereby renewable energies replaced lignite (25.4%) as the most important energy sources for electricity generation."[56]

---

54 Ibid. p. 667
55 Der neue Fischer Weltalmanach 2016, p. 667, with figure
56 Ibid., p. 669, with figure on the second next page

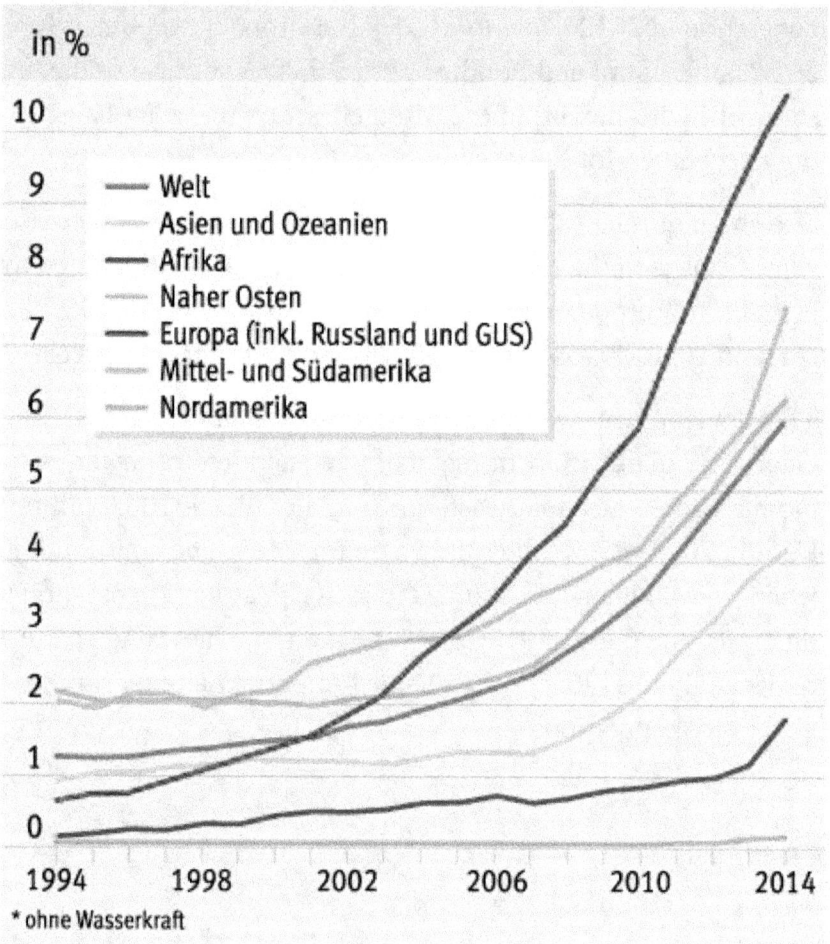

Anteil erneuerbarer Energieträger* an Stromerzeugung nach Regionen

* ohne Wasserkraft

Source: BP 2015

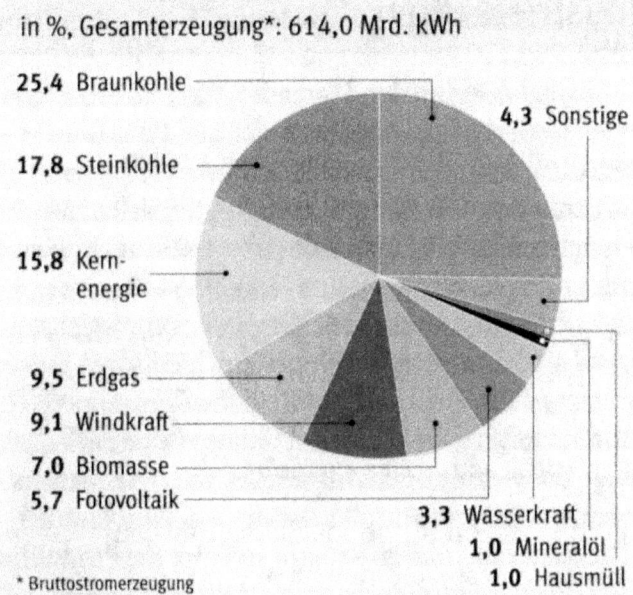

Source: AG Energiebilanzen 2015

4. Climate Development in the Last Century

Climate changes primarily affect the general circulation of the atmosphere, air pressure, temperature and precipitation. Various feedback effects have been discovered over the last few decades. In addition to natural climate changes, for example due to the variable influences of the sun and volcanic eruptions, there are climate changes caused by humans, especially "through energy supply, waste gases, increased carbon dioxide, trace gases and changes due to the destruction of vegetation."[57]

Spatially limited impacts on the atmosphere often cause severe damage, such as waste heat, acid rain, smog, photochemical oxidants and air pollution. Multiplying these effects on a global scale results in much greater problems, the anthropogenic causes of which are difficult to prove, "as is the fact that, since industrialisation about 100 years ago, the global temperature of the atmosphere at ground level has increased by 0.7°C, most significantly in the last ten years."[58]

Human activities during the course of industrialisation have changed the composition of the atmosphere to such an extent that a threat to the survival of life on earth has gradually developed and is continuing to grow. The atmospheric concentration of greenhouse gases and the average global temperature are subject

---

57 DIE ZEIT: Das Lexikon in 20 Bänden (The Lexicon in 20 Volumes), Volume 08, p. 55
58 Ibid.

to natural fluctuations that are increasingly being overlapped by the impact of human activities. Since the beginning of industrialisation, this has led to an increase in greenhouse gases and to global warming..."[59]

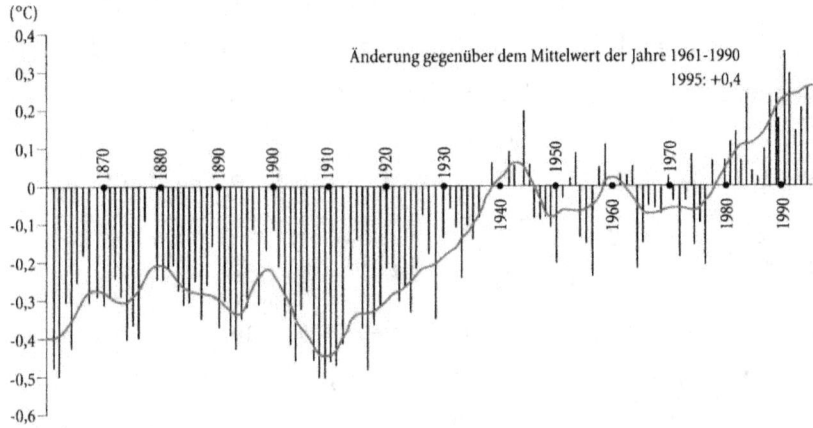

Source: Hadley-Centre 1996

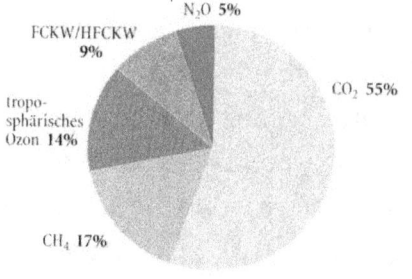

"The contribution of individual greenhouse gases to global warming is derived from the respective increases in concentration and the capacity of the earth to absorb heat radiation...
We have known since recently that aerosols are produced when

Source: Intergovernmental Panel on Climate Change (IPCC), 1995

---

59   Der Fischer Weltalmanach 1997, p. 1120f, with figure

burning fossil fuels (coal, oil and natural gas) and biomass, which have a potentially cooling effect on the climate through enhanced reflection of solar radiation. The climate impact of increasing greenhouse gas concentrations are thus offset by about 20%"[60].

This development resulted in policy measures in 1992: A Framework Convention on Climate Change was signed by 159 countries, which came into force on 21/03/1994 and was ratified by these countries until 1996. The first Conference of the Parties to the Framework Convention on Climate Change took place from 28/03 to 07/04/1995 in Berlin and determined that limitation and reduction targets should be adopted for defined time-frames by the third Conference of the Parties in 1997 in Japan.[61]

Toward the end of 1997, further developed and specific commitments to combat climate change were made public in **Kyoto** (Japan) by the adoption of the "Kyoto Protocol" on the occasion of the third Conference of the Parties to the Framework Convention on Climate Change.[62]

Some years later, it turned out that the ten years from 1995 to 2005, with the exception of 1996 and 2000, were the hottest since records began. "Regionally, **climate extremes** like those experienced in the previous years were very unevenly distributed: **There were heat waves** in Australia (warmest year since records began in 1910), India, Pakistan and Bangladesh (May/June, record temperatures of 45-50°C, delay in the south-west monsoon, at least 400 deaths in India), in the south-west of the USA

---

60   Ibid. p. 1123 with figure
61   Ibid. p. 1125
62   Der Fischer Weltalmanach 2004, p. 1318

(beginning of July), in Central Canada (the warmest and wettest summer to date), in China (one of the warmest summers since 1951), the South of Europe and Northern Africa (July, temperatures in Algeria of up to 50°C, several deaths). **Cold snaps** were observed in the Balkans (at the beginning of February), in Morocco (January, temperatures as low as -14°C) and in the East Asian region (December, Japan and Korea). The long years of **drought** along the Horn of Africa (southern Somalia, east Kenya, south-east Ethiopia and north-east Tanzania) continued - 11 million people in this region were in danger of starvation. Among other areas, drought also affected southern Africa (5 million people starving in Malawi), Western Europe (worst droughts in Spain and Portugal since the 1940s), South Brazil (December, corn and soybean harvest greatly affected) and the Amazon Basin (lowest water levels for 60 years). Torrential rains and **floods** during the monsoon period (June-September) affected 20 million people in West and South India resulting in 800 deaths. On the 27/07/2005, 944 mm of rain fell in Mumbai - more than ever before in a single day. Approximately 450 people fell victim to the floods. There was further flooding from October to December in South India (300 deaths), Thailand (52 dead) and Vietnam (69 dead), in the third week of June in southern China (170 dead), in Eastern Europe (66 dead in Romania alone) from May to August, in Costa Rica and Panama (35,000 refugees) in January and in Colombia and Venezuela (80 dead) in February."[63]

---

63   Der Fischer Weltalmanach 2007, p. 710f, with figure

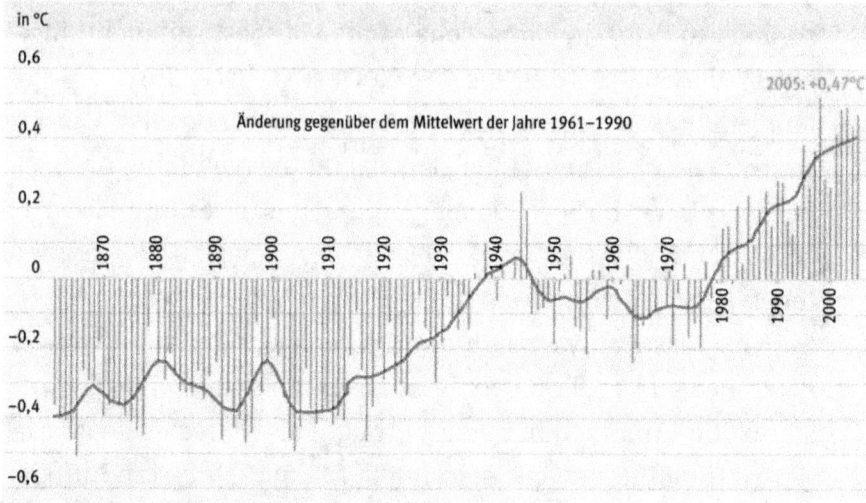

Source: WMO

"The **rise in sea levels** of 10-20 cm observed over the last one hundred years is probably largely due to global warming. This process may be accelerated by the melting of the polar ice caps. The sea ice in the Arctic is now only half as thick as it was 50 years ago, its spread in 2005 was well below the long term average for the fourth year in a row (-20% compared to the period from 1974-2004). By 2100, the "Intergovernmental Panel on Climate Change (IPCC)" submitted at UN level expects a further increase in the global mean temperature of 1.4 to 5.8°C and a rise in the sea level of 9 cm to 88 cm, if no appropriate measures are taken. New calculation models published at the end of September 2005, in which 15 research groups worldwide were involved, showed a temperature increase of 4°C if greenhouse gas emissions continued to increase as before, and 2.5°C if at least the requirements of the Kyoto Protocol are met.

According to new studies by US researchers published in the scientific journal 'Science' at the end of March 2006, the ice sheets in Greenland and the Antarctic could be melting much faster than previously estimated: The Arctic summer in 2100 may therefore be as warm as it was almost 130,000 years ago. The sea level then was six meters higher than it is today. As the destruction of the ice caps and the subsequent rise in sea levels is occurring with a time delay, this process will become irreversible sometime in the second half of the 21st Century. The researchers believe that the sea levels will have risen by four to six meters by 2100, if the greenhouse gas emissions are not reduced quickly and permanently.

A highly visible consequence of global warming is the accelerated recession of Alpine glaciers. The melt began in the middle of the 19th Century, when the Alpine glaciers still had a volume of 200 km^3. In 2000, this still came to 75 km^3 and had decreased to only 68 km^3 by 2005. From 1985 and 2000 alone, the Alpine glaciers lost 20% of their surface area and one quarter of their volume. A melt of this magnitude was actually only expected by 2025. The rate of surface loss has doubled from 1% annually from 1973 to 1985 to 2% p.a. today."[64]

Climate change has thus been developing at a previously unpredicted tempo and can no longer be denied in consideration of La Niña (cooling weather phenomenon) and El Niño (warm

---

64   Ibid. p. 711f

weather phenomenon).[65]

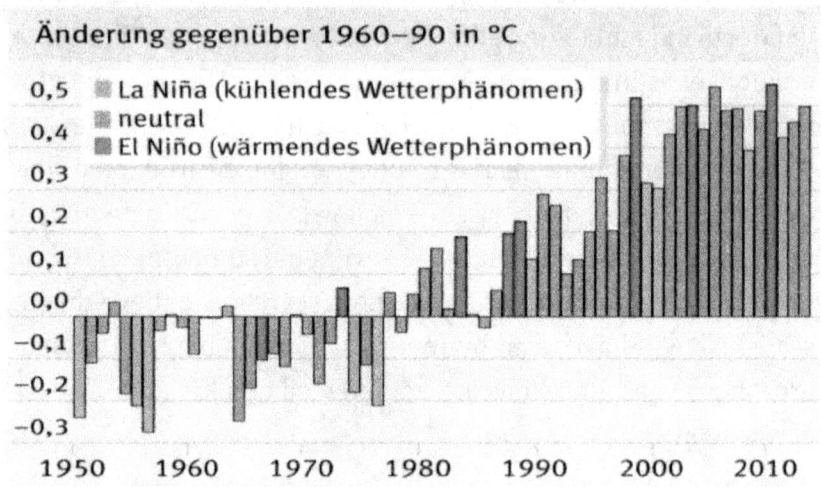

Source: WMO 2014

"According to the World Meteorological Organization (WMO), 2013 (together with 2007) was the sixth warmest year since records began in 1850. The temperature on the surface of the earth was on average by 0.5°C above the average for the reference period from 1961-90 of 14.0°C. It was particularly hot in the southern hemisphere: In Australia, 2013 was the hottest year since records began and in Argentina, the second hottest recorded.

Since the beginning of the 20th Century, the world climate has heated up by approx. 0.75°C, with 13 of the 14 warmest years in the 21st Century. Each of the last three decades has been warmer

---

65   Der Fischer Weltalmanach 2015, p. 693, with figure

than the previous one. The decade from 2001-2010 was the warmest to date on all continents, with the temperature being around 0.46°C higher than in the reference period of 1960-90. Reconstruction of the climate in the past, based on tree rings, corals, ice cores and sediments, has shown that the average temperature in the northern hemisphere, at least in the last 1400 years, has never been as high as it is today."[66]

The causes of climate change can now be quantified very accurately. "The Fifth Progress Report of the IPCC (Climate Change 2014) summed up the findings of global climate research with the following words: "The warming of the climate system is clear, and the like of the changes since the 1950s have not been seen for decades or millennia. The atmosphere and the oceans have warmed up, the snow and ice have receded, the sea level has risen and the concentration of greenhouse gases has increased." The main cause of global warming in recent decades is the accelerated **emission of greenhouse gases** by humans. Accounting for around three quarters of the total emissions, **carbon dioxide** (CO_2) is the most important greenhouse gas. It mainly comes from the combustion of fossil fuels and to a lesser extent the clearing of forests, which bind CO_2 from the air as they grow and thus act as "CO_2 sinks". After the forests, the oceans are the most important 'sinks' in the CO_2 cycle". Their effectiveness initially increased with increasing emissions. According to the calculations of the Global Carbon Project, between 1958 and 2010 56% of anthropogenic CO_2 emissions were "buffered" in this way. However, the buffer capacity of the oceans and forests has

---

66   Ibid. p. 693f

decreased in recent times. Of a tonne of CO_2 released in 2010, only half is absorbed directly by the oceans (24%) and biosphere (26%). The remaining 50% leads to an accelerated increase in the CO_2 concentration in the atmosphere.

Other greenhouse gases include methane (CH_4, mainly from livestock farming, petroleum and natural gas production and rice cultivation), nitrous oxide (laughing gas, N_2O, mainly from over-fertilised soil) and fluorinated gases. These include, among others, perfluorocarbons (PFCs) and sulphur hexafluoride (SF_6), mainly originating from industrial processes, as well as chlorofluoro-carbons (CFCs) and partially halogenated hydrocarbons (HCFCs), which are used as coolants and solvents. The sum of all greenhouse gas emissions is expressed in CO_2 equivalents (CO_2e).

N_2O, CFCs and - to a lesser extent - HCFCs also cause the depletion of the ozone layer ("ozone hole") in the stratosphere. This is above the troposphere, the lowest layer of atmosphere, in which climate and weather events take place. Based on the UN Convention for the protection of the ozone layer (Montreal Protocol of 1987), production and use of CFCs and HCFCs is decreasing worldwide, and with this the ozone layer is slowly recovering.

According to the most recent IPCC report, the concentrations of CO_2, methane and nitrous oxide in the atmosphere now far exceed the highest concentrations of the last 800,000 years, which we know from the ice cores. The increase over the last century is greater than at any time in the last 22,000 years. And they continue to rise: 1970-2000 by 1.3% annually, 2000-10 by 2.2%.

The sum of all anthropogenic greenhouse gas emissions from 2000-2010 was higher than ever before; in 2010 it came to 49 billion tonnes of CO_2e.

Around half of the cumulative global CO_2 emissions from 1750-2010 have occurred over the last 40 years. In 1970, the accumulated CO_2 emissions from the combustion of fossil fuels, the production of cement and gas flaring was around 420 Gt according to the IPCC; in 2010 this was around 1300 Gt. The accumulated CO_2 emissions from forestry and other land use not including agriculture since 1750 increased from around 490 Gt in 1970 to around 680 Gt in 2010. The economic impact of the global financial and economic crisis of 2007/2008 also reduced emissions for a short period of time. The People's Republic of China has for years been the country with the highest - and further increasing - CO_2 emissions."[67]

In view of this development and in response to the Fukushima disaster, in 2011 the German Bundestag adopted a comprehensive legislative package on the energy transition. "The complete package includes the phasing out of nuclear energy by 2022, the accelerated expansion of renewable energies, the development of electricity networks and storage capacities, greater heating energy savings in the building sector and the introduction of electric mobility."[68]

---

67   Ibid. p. 694f with figure
68   Ibid. p. 700 with figure

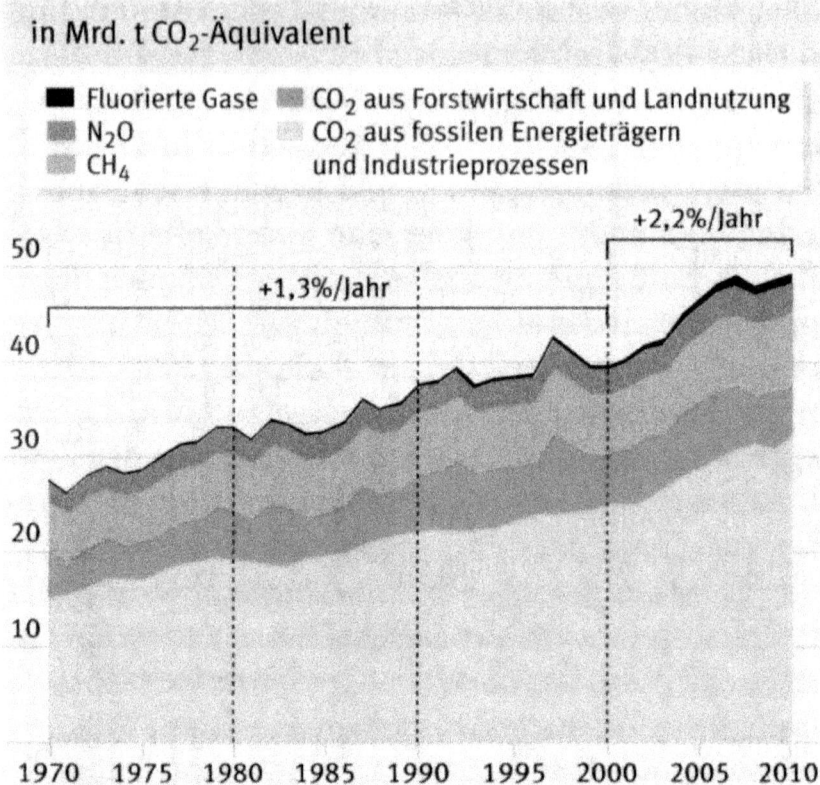

Source: IPCC 2014

Deutschland: Klimaschutzziele

Gruppe	Ziel (Bezugsjahr 1990)			
	2020	2030	2040	2050
Treibhausgasemissionen	−40%	−55%	−70%	−80 bis −95%
Erneuerbare Energien: Anteil am Bruttoendenergieverbrauch	18%	30%	45%	60%
Erneuerbare Energien: Anteil an der Stromerzeugung	35%	50%	65%	80%
Primärenergieverbrauch (gegenüber 2008)	−20%			−50%
Stromverbrauch (gegenüber 2008)	−10%			−25%
Endenergieverbrauch im Verkehr (gegenüber 2005)	−10%			−40%

Source: BMWi/BMU 2011

On 03 August 2015, Deutschlandfunk radio broadcast the following headline: "**US President Obama wants to reduce CO_2 emissions from power plants in the United States with binding targets.**" followed by the details:

"He presented a climate protection plan at the White House in Washington to reduce emissions by one third by 2030 compared to 2005. Obama stressed that climate change is the greatest threat to the future of humanity - which only has one home and one planet. The Environmental Protection Agency (EPA) submitted the main features of the provision one year earlier. EPA Head McCarthy called the intended target "reasonable" and "achievable" today. According to the White House, round 1,000 power plants are affected in the United States, including 600 coal-fired power plants.
Federal Environment Minister Hendricks explained that she welcomed the fact that the United States was facing the challenge of climate change. The new plan is an important signal for the Climate Conference in Paris at the end of the year."[69]

The following day, according to Deutschlandfunk radio, Obama called for increased efforts at the UN.

Headline: "**US President Obama has called for increased efforts to protect the climate with UN Secretary-General Ban Ki Moon.**"

---

69 http://www.deutschlandfunk.de/programmvorschau.281.de.html?drbm:date=03.08.2015

US President Obama with Ban Ki Moon in the White House Oval Office (picture alliance / dpa / EPA/Dennis Brack / POOL)

Details: "The United Nations could contribute to the fight against climate change, Obama said after the meeting at the White House. The UN must increase pressure on other countries to also make efforts to reduce harmful emissions. The day before, the US President presented a plan, according to which American power plants are to reduce their emissions of greenhouse gases by almost one third by 2030. Ban welcomed Obama's proposal. The current generation is the last one that will be able to tackle the phenomenon of climate change.

Several candidates for the Republican presidential candidacy have criticised the plans and warned against the loss of jobs and rising electricity costs."[70]

---

70 http://www.deutschlandfunk.de/programmvorschau.281.de.html?drbm:date=04.08.2015, with figure

The weekly newspaper "DIE ZEIT" addresses the same topic in its Issue No. 32 of 06 August 2015 on page 23: **"Gemeinsam schnell die Welt retten" (Saving the World Together Quickly)**.

Subtitle: "Die Präsidenten Barack Obama und Xi Jinping setzen ihren Ländern echte Klimaziele. Eine einmalige Chance" (Presidents Barack Obama and Xi Jinping set real climate targets for their countries. A unique opportunity) by Claus Hecking. This shows that the climate goals of the USA and China are not as ambitious as they seem. Thanks to its reserves of shale gas, to generate electricity the USA burns "cheap natural gas instead of coal... The CO_2 emissions per kilowatt hour are only about half as high. In 2008, coal-fired power plants were still producing almost of half of US power, this year it will be only one third. Obama has wisely chosen 2005 as the reference year for his climate targets, not 2015. Since then, the energy industry has reduced its CO_2 emissions by about 16 percent. Half the work is already done. Beijing also promise that China's CO_2 emissions will decrease after 2030, which sounds more ambitious than it actually is. The national coal consumption already decreased by almost 3 percent last year, even though total energy consumption increased by a good 2 percent. Since China is replacing old coal-fired power stations with more efficient nuclear reactors and building giant wind and solar parks, the CO_2 emissions have already fallen since 2014."[71]

---

71 Hecking, Claus: Gemeinsam schnell die Welt retten (Saving The world Together Quickly) in: DIE ZEIT No. 32 2015, Hamburg 06/08/2015, p. 23

5. Future Perspectives

Since the beginning of the 21st Century, numerous sensational and sometimes catastrophic weather phenomena have led to crop failures, storm damage, flood disasters and drought. Even those not directly affected are noticing this, as their insurance is becoming more expensive.

Most strikingly, this development has become apparent with the insurers of the insurance companies, re-insurance companies such as Munich Re, the Munich re-insurer, because the insured losses are adding up there. Therefore, it is not surprising that these institutions are joining forces with politicians and industry to find solutions to the climate problem.

Thus, attentive was drawn to a development around the Mediterranean Sea and especially in desert areas: the development of the DESERTEC Concept (2003 to 2007).

"The DESERTEC Concept was developed by an international network of politicians, scientists and economists. From this Trans-Mediterranean Renewable Energy Cooperation (abbreviation: TREC) Network came the later DESERTEC Foundation. The physicist Dr. Gerhard Knies and Prince Hassan bin Talal of Jordan, the then President of the Club of Rome, were the driving forces behind the foundation and development of the network. The research facilities for renewable energies of the governments of Morocco (CDER), Algeria (NEAL), Libya (CSES), Egypt (NREA), Jordan (NERC) and Yemen (Universities of Sana'a and

Aden), as well as the German Aerospace Center (DLR) played decisive roles in the development of the DESERTEC Concept. The underlying studies on DESERTEC were directed by DLR researcher Dr Franz Trieb. The studies were financed by the German Ministry for the Environment (BMU), which at that time was led by federal ministers Juergen Trittin and later Sigmar Gabriel."[72]

The DESERTEC Concept plans to generate clean electricity in the desert and transport it to the centres of consumption up to 3,000 km away.

"The deserts of the earth receive more energy from the sun in six hours than mankind consumes in one year. The DESERTEC Concept stands for the large-scale production of solar and wind energy in the desert regions of the world, combined with an intelligent mix of photovoltaics, hydropower, biomass and geothermal energy. By using these renewable energies within a trans-national network, enough clean electricity can be generated to supply the whole of humanity".[73]

The required technologies are already available and are used commercially all over the world.

"DESERTEC is technology neutral. The DESERTEC Concept integrates **all types of renewable energies** in a trans-national super grid. However, a key technology in the DESERTEC Concept is solar thermal energy. As a controllable source of

---

[72] http://www.desertec.org/de/globale-mission/meilensteine/
[73] http://www.desertec.org/de/konzept/

energy, it is in the position of being able to balance fluctuations in wind and photovoltaic energy."[74]

Examples for the use of these technologies include "the Andasol Solar Thermal Power Plant in Andalusia (Spain) and Solar Park in the Mojave Desert in California (USA).

In the case of solar thermal electricity production, solar energy is concentrated by mirrors in order to heat water. The resulting steam is used to drive a conventional power turbine. Because large quantities of heat energy, in contrast to electricity, are technically easier to store with a low loss rate, these power plants can deliver power on demand - even after the sun has gone down. A large proportion of clean and controllable energy in the energy mix stabilises the network and allows for more efficient use of fluctuating energy sources such as wind and photovoltaics.

In regions with constant, high solar radiation, solar thermal power plants can be used particularly efficiently. For this reason, desert regions are ideal production locations."[75]

---

74 Ibid.
75 Ibid., p. 26, with figure on the next page

Parabolic trough   Fresnel collector

Solar tower

"Clean power from the desert can be transported over long distances via high-voltage direct current lines. 90% of humanity could theoretically be supplied with clean electricity from the desert, because they live within 3000 km of a desert. At only 3% per 1000 kilometres, the loss rate is relatively low - the location advantages of solar energy plants in deserts more than balance these line losses out.

China in particular has already gained experience of the use of high-voltage direct current transmission lines (HVDC transmission lines), such as the 1418km long HVDC line between Yunnan and Guangdong, for example."[76]

In 2008, the Solar Plan for the Union for the Mediterranean (UfM) began. The Mediterranean Solar Plan has set itself the target of implementing renewable energy projects with a total capacity of 20 gigawatts by 2020.

Following this, on 20 January 2009, the DESERTEC Foundation was founded as a charitable foundation, "to facilitate the implementation of the global DESERTEC Concept of "Clean Power from Deserts" worldwide. The founding members of the DESERTEC Foundation are the German association, Club of Rome e.V., members of the TREC network of scientists as well as dedicated private investors and long-standing supporters of the DESERTEC idea."[77]

From 2009 to 2014, commercial enterprises examined the efficiency and feasibility of the DESERTEC vision with a positive outcome. In 2012, the consulting firm Dii GmbH, initially engaged for this purpose for three years, had already highlighted the economic feasibility and the unique benefits of an EUMENA region power network. The 2013 "Getting Started" report finally confirmed both its economic attractiveness and feasibility. The technical and energy-economic prerequisites for generating electricity from renewable energy under competitive conditions

---

76   Ibid.
77   http://www.desertec.org/de/globale-mission/meilensteine/

are already in place and transport both within the MENA region as well as between the EU and MENA is already economically attractive today. Thus, large-scale CSP, wind and PV power plants could generate electricity at a much lower cost than power stations that operate using oil. The Mediterranean region is, from an energy policy perspective, to be understood as the centre rather than the boundary in the long term. How the expansion of renewable energies throughout the entire EUMENA region could be made possible from the perspective of industry is described in clear recommendations for action. After completing its task, Dii GmbH is to be operated as a consulting firm by three companies from 2015."[78]

Wind power is also being further developed in the North Sea again. With the headline "**Offshore-Branche schöpft wieder Hoffnung**" (**Offshore sector generating hope again**) and the subtitle "Nach Jahren der Krise zeichnet sich eine Wende ab – Siemens investiert Millionen – ABB nimmt Netzanbindung in Betrieb" (After years of crisis, there are signs of a turnaround - Siemens invests millions - ABB takes network connection into operation), Ralf E. Krueger and Christine Schultze describe a promising development in the wind power industry in the North Sea in an article in the newspaper, Rhein-Neckar-Zeitung, of 08/09 August 2015.[79]

---

78 Ibid.
79 Krüger, Ralf E., Schultze, Christine: Offshore-Branche schöpft wieder Hoffnung (Offshore sector generating hope again), in Rhein-Neckar-Zeitung / No. 181 of 08/09 August 2015, p. 22

The Siemens electrical and electronics group is building a new plant for offshore wind turbines in Cuxhaven.

"In the first half of the year alone, 422 new offshore wind power plants with a capacity of 1765.3 Megawatt (MW) joined the grid, according to the German Wind-guard. At sea, by the end of June, 668 plants with a capacity of 2777, 8 MW of electricity were already feeding in. Europe is by far the largest offshore market in the world with a good 8000 megawatts of installed power.

This week, the offshore network connection Dolwin1, built by ABB, was put into operation and handed over to the German-Dutch transmission system operator Tennet. The 800-Megawatt DC power plant connects offshore wind farms in the Dolwin Cluster about 75 kilometres off the German coast to the country's transmission network."[80]

Dolwin is the grid connection built by ABB, which initially brings electricity from some 160 offshore wind turbines onshore. Photo: ABB

---

80   Ibid., with figure

From 2009 to 2013, the framework conditions for wind energy connection were really neglected in terms of policy. This has now changed. For Siemens the new location is, therefore, an important step toward more cost-effective production.

The costs for the construction of wind farms may still be high - but the Munich company is working on the industrialisation of the business... This should also contribute to better logistics. Thanks to the well-developed port facility in Cuxhaven, heavy components can be loaded directly onto transport ships."[81]

In the course of expanding the wind park and photovoltaic facilities, it became increasingly clear that too much electricity is often generated when the wind is very strong or the sun is shining, which on the other hand leads too high-pressure weather conditions with lulls. At night, electricity from photovoltaic systems is not available. Here, storage systems are required, which can take over in the event of power shortages and deliver the stored electricity.

The most well-known storage systems are reservoirs which are filled with water from the valley if there is an excess of power and released if there is too little power, thus driving turbines to generate power. However, the capacity of these reservoirs is not nearly enough.

A further possibility for storing electricity is offer by accumulators, but these are far too expensive for mass capacities. A third possibility for storing excess electricity is electrolysis. Using this, water is split into its component parts of hydrogen and

---

81   Ibid.

oxygen. The gases are then stored in suitable tanks and, if there is too little power, combined in fuel cells as water, whereby they release combustion energy in the form of electric current. This technology has been tested and is currently being further developed for large-scale use.

The state of development is described by Katja Scherer in an article in the weekly newspaper DIE ZEIT No 18 of 29 April 2015 on page 31.[82]

Under the headline "**Rein ins Rohr**" (Into the Pipe), with the subtitle "Wenn die Sonne scheint und der Wind weht, wird zuviel Strom produziert, sonst zu wenig. Helfen könnte das Gasnetz, wenn man es in einen Speicher verwandelt" (When the sun is shining and the wind is blowing, too much electricity is generated, otherwise too little. The gas network could help if it were turned into a storage system), the article concludes that: "It will have to be possible to store **50** terawatt hours of electricity from renewable energies by 2050 - three times as much as 2020. The gas grid could help with this."[83]

A container in the industrial district of Frankfurt am Main contains a test plant for this electrolysis technology. The heart of the facility is the PEM electrolyser, a Proton Exchange Membrane. "It allows hydrogen to be extracted from water using electricity, i.e. transforming electrical energy into chemically bound energy. The hydrogen gas thus becomes a form of

---

[82] Scherer, Katja: Rein ins Rohr (Into the Pipe), in: DIE ZEIT No. 18, Hamburg 2015, p. 31
[83] Ibid.

electricity storage. This procedure is called power-to-gas."[84]

This makes it possible to also store large quantities of excess renewable energy. According to Thüga's calculations, the storage requirement for renewable energies in 2020 will be 17 terawatt hours and will grow to about 50 terawatt hours by 2050. In order for the energy turnaround to work, Germany needs long term procedures to store the energy generated from renewable sources. The existing gas network of suppliers should provide help: Its annual capacity according to Thüga is four times greater than the requirement in 2050. With the help of power-to-gas, it could soak up energy that would normally go unused like a sponge - and release it again when there is too little in the power grid."[85]

However, there are strict conditions for feeding into the Frankfurt natural gas network: The proportion of hydrogen within the gas network may not be greater than two percent according to law. This is to prevent a natural gas filling station somewhere in Frankfurt from suddenly blowing up, as hydrogen is highly flammable.

Another option is the power-to-gas process, which further processing hydrogen to methane. This has similar chemical properties to conventional gas and can therefore be fed into the gas network without restriction."[86]

"Natural gas vehicles could also be fuelled by the methane obtained from renewable energy ... practice testing is already

---

84 Ibid.
85 Ibid.
86 Ibid.

underway: Since 2013, the car manufacturer Audi has had a pilot plant in Werlte, Lower Saxony."[87]

Making gas from electricity: This is the idea for the new technology. Here, a demonstration plant on the premises of Mainova AG at Frankfurt.

Alternative Drive Concepts are, however using the re-conversion of hydrogen into drive energy with a much higher level of efficiency, for example the purely electrically driven fuel cell car from Toyota, which is available from September of this year (2015). However, there are far too few hydrogen filling stations for private customers, so it is more suited to customers such as taxi companies and city works. On a larger scale, the fuel cell technology still appears to be too expensive.

In the United Arab Emirates, the more extensive development of an ecocity has come about already. Masdar City is a city-building

---

87   Ibid., with figure

project in the Emirate of Abu Dhabi that started in 2008.[88]

The city currently under construction will be fully supplied by renewable energy. The water desalination plants are to survive on solar energy. The total energy consumption of the city only comes to a quarter of the country's usual per capita consumption. The energy supply is to be completely free of carbon dioxide production.

"Masdar is being built about 30 kilometres east of the capital city of Abu Dhabi, to the west of and adjacent to Abu Dhabi International Airport. The ambitious project in an area of six square kilometres is being designed for 47,500 inhabitants and around 1500 companies and institutes in the ecological sector, with no point in the city area being more than 200 meters from a public transport stop. The initiative is spearheaded by the Abu Dhabi Future Energy Company (ADFEC) and Sheikh Mohammed bin Zayid Al Nahyan. Initiated in 2006, the plan was for the first inhabitants to move in from 2016… in the spring of 2010, however, delays and financial problems were reported by various media. The construction works have lost pace and determination, as the new completion date for the entire project is now 2025".[89]

Masdar will also be home to a new university, the first in the world, "dedicated completely to the complex of ecological sustainability on the basis of renewable energies. The first university facilities have already been moved into since 2009, one third of the students will live in the Masdar area and be involved

---

88 http://masdar.ae/
89 https://de.wikipedia.org/wiki/Masdar

in city planning and construction within the context of their courses of study. It is also expected that companies based in and around Masdar and their institutions will gain new experience during the course of the construction projects, will have to apply special technological procedures or generate new ecologically usable knowledge that they will be able to sell on the growing world market for sustainable systems."[90]

The energy supply will be ensured by its own solar power plant and a wind park. There will be no fossil-fuelled vehicles in Masdar; they have to remain outside the city. Passenger transport there will be in the form of electrically operated public transport.

"Seamless transport in the model city is to be provided by various coordinated forms of public transport, each of which is to be assigned to a level. In the Masdar underground and two other Abu Dhabi city districts, local so-called Personal Rapid Transit Networks (PRT Networks) are to be installed by the Dutch company 2getthere. This involves electrically motorised individual transport, whereby the user gets into an automated cab without having to wait and goes to the destination specified by him... Since August 2011, the system has been in testing in Masdar City using ten cabs, with use for separate freight transport included in the programme. The users gets into or out of the cabs at the stops secured by separating doors and travel along ground level leading strips at up to 40 km/h along the transport deck.

Therefore, Masdar will be the first city in the world to use a PRT network for a car-free city. No cars are allowed on the (ground

---

90   Ibid.

level) streets, the so-called "podium level". These are only for pedestrians and cyclists. At a higher level, an elevated railway (*Light Rail Transit*, LRT) is planned, which will connect Masdar with other parts of the city and the airport. Furthermore, a regional train is also planned below the PRT level."[91]

Station for driver-less cab vehicles

"According to the initial plans, Masdar City was to be completed by 2016; since January 2010 however, the final completion date has been pushed back at least until 2025. Only the Masdar City urban core sub-project is to be up and running by 2016. The official reason given for this is the consideration of further new technologies. Since May 2009, work on the foundations of the headquarters of the ecocity, The Masdar Institute of Science and Technology, opened at the beginning of the 2010/2011 semester with 170 hand-picked postgraduate students. Apart from this, there has been hardly any progress with construction due to the financial crisis. Because of this, the tenders for the construction of apartments and offices, for example, have not been awarded yet.

It is said within the project environment that the major part of the

---

91   Ibid., with figure

urban planning project has been suspended.

The main obstacle is the lack of planning security due to the country's autocratic leadership. Agreements by the ruling family of the Emir can be revoked at any time."[92]

In Sven Plöger's book "GUTE AUSSICHTEN FÜR MORGEN" (Good Prospects for Tomorrow) with the subtitle "Wie wir den Klimawandel für uns nutzen können" (How we can use climate change to our advantage)[93], which is well worth reading, this problem has not yet been addressed. But development continues:

"The **International Renewable Energy Agency** (abbreviation: **IRENA**) is an international intergovernmental organisation that aims to promote the comprehensive and sustainable use of renewable energies all over the world. Its head office will, in the future, be in the ecocity, Masdar City, in the United Arab Emirates.

As of January 2015, 138 states and the European Union are members of IRENA. A further 35 states have applied for membership... The Statute of the organisation entered into force on 08 July 2010, on thirtieth day after the twenty-fifth ratification (Art. XIX lit. D of the Statute...).

Since 03 April 2011, the Kenyan, Adnan Z. Amin has been the General Director of IRENA... Amin had acted as interim director for some months before this, after his predecessor Hélène Pelosse unexpectedly resigned from the position after less than six months

---

92 Ibid.
93 Plöger, Sven: GUTE AUSSICHTEN FÜR MORGEN (Good Prospects for Tomorrow) Frankfurt/Main and Munich, 2nd Edition 2010, p. 295f

in office.

Together with the IEA, IRENA places extraordinary importance on energy issues in general as well as on topics concerning renewable energies in particular."[94]

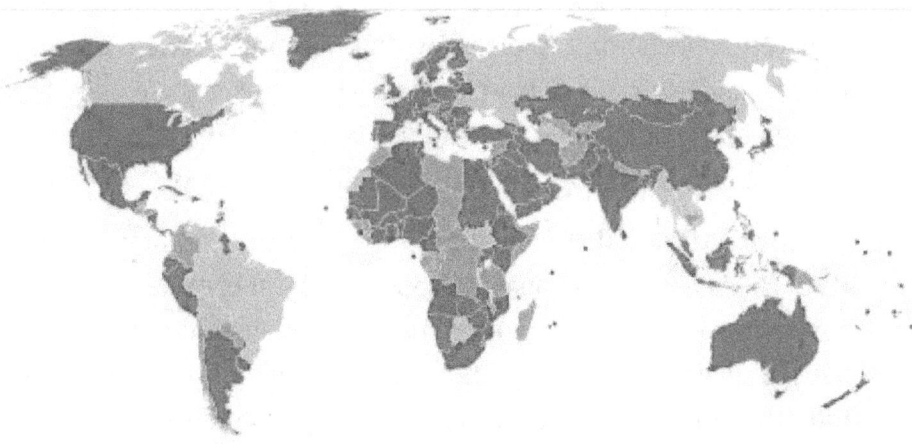

Blue: Countries that have signed the agreement on the International Renewable Energy Agency
Green: Countries that have signed and ratified the Agreement, as of 17 January 2015

"IRENA was founded on 26 January 2009 upon signing of the Statute by 75 states in Bonn... At the first meeting of the Preparatory Commission, the temporary body by which IRENA was embodied up to the 25th Ratification of the Statute, the signatory states agreed on the rules and the selection procedure for the provisional Director General and the seat of IRENA.

In addition, the members were called upon to submit nominations

---

94 https://de.wikipedia.org/wiki/Internationale_Organisation_für_erneuerbare_Energien

and applications for the seat and the Director General. At the second meeting of the Preparatory Commission on 29/30 June 2009 in Sharm El Sheikh (Egypt), both the seat and the Director General were decided on, after which the seat of IRENA was to be in Abu Dhabi, the seat of a Centre for Innovation and Technology was to be established in Bonn and a Liaison and Contact Office to the United Nations in the field of energy and other international institutions in Vienna. A Seat, Director General and Management Committee was launched, which was to prepare the contents of the second meeting by then. Other points on the agenda for Egypt include the adoption of a work programme, the budget and the financial and staff regulations for IRENA that were to apply for the transitional phase from 2009 to 2010.

Three years after being established, the organisation is gaining momentum due to the rapid growth in membership. In various analyses and papers, IRENA has compiled comprehensive data material on the global expansion of renewable energies. In addition to the overview of the cost situation of renewable energies in 2012, the Global Atlas is of particular significance and is to be expanded and supplemented over the next few years. This provides information for investors on the potential of the renewable energy sources in various countries. There is already a strategy for Capacity Building, i.e. the transfer of information, education and training regarding renewable energies."[95]

Policy is therefore obliged to prevent too much global warming. This was also clearly reflected at the last G7 Summit, which took place in Garmisch-Partenkirchen in the municipality of Krün at

---

95   Ibid., with figure

Elmau Castle.

On 07/08 June 2015, the Heads of State for the United States, Japan, Canada and the four European countries, France, Great Britain, Germany and Italy, meet to discuss the most pressing problems in world events.

The weekly newspaper, DIE ZEIT writes in its commentary of 08 June 2015 at 20:26 under the headline "**Elmau-Gipfel Die G7 allein können es nicht richten**" **(Elmau G7 Summit cannot do it alone)**[96] in the subtitle:

"Die Lenker der sieben großen Industrieländer treffen in Elmau viele, zum Teil vage Vereinbarungen. Einlösen können sie diese jedoch nur mit anderen. Ein Kommentar von Carsten Luther. Schloss Elmau." (The leaders of the seven major industrial countries reached many, somewhat vague agreements in Elmau. But they can only honour these together. A commentary by Carsten Luther. Elmau Castle.).[97]

---

96 http://www.zeit.de/politik/deutschland/2015-06/g7-ergebnisse-kommentar
97 Ibid., with figure

The G7 participants led by Angela Merkel on a walk through the flower meadow in front of Elmau Castle © Christian Hartmann/Reuters

"Other forums of the club of seven have run their course, in which China, emerging countries and Russia, excluded again in Elmau, were represented. How to respond to climate change, how to progress and finance the development goals of the United Nations, how to make steps toward peace in Ukraine, Syria and the Middle East and how to suppress terrorism? How to secure sustainable growth that is not at the expense of people and the environment? Almost all of the problems facing the world were addressed at this summit, all of which will continue to be discussed - along with others if anything changes. More concrete responses to global challenges than those at this cheerful meeting in Elmau are more to be expected within the context of the G 20 and other talks."[98]

---

98   Ibid.

"For example in the case of climate change: It does not mean much that the Leaders of the G7 states have only confirmed a willingness to limit global warming to two degrees in comparison to the pre-industrial era or if they endeavour to phase out fossil fuel entirely "over the course of the century". It is only a stage, the lowest common denominator. Laying the groundwork for the UN Climate Conference in Paris in December...

Rather, there is something that can done if the seven promise money: along with the commitment to full the previously planned climate protection fund worth billions for developing countries - one can only work with this if is really financed...

And yet, this format is not an anachronism in a time when all problems are global. The session held in Elmau is not a group that determines the world order alone, but they still have weight. If they want. What the seven agree here in partnership-based dialogue can be pushed forward with a common voice within the more weighty G-20 or UN mechanisms; they can convince and act as pioneers in regard to many questions."[99]

Climate protection is clearly no longer just a buzzword but rather increasingly concerns the world population in the face of increasingly violent weather conditions and environmental disasters. The consequences of climate change are now well documented. "Due to the long retention period for greenhouse gases in the atmosphere (CH_4: 12 years, CO_2: 120, SF_6: 3200) the climate will continue to heat up over the coming decades.

---

99   Ibid.

The rate of temperature increase and the associated climate impacts will depend on whether and how much emissions can be reduced by humans. For its most recent progress report, published in 2013, the IPCC calculated four scenarios, each with greenhouse gas concentrations developing differently. Depending on the scenario, warming in the 21st Century in the best case will be 0.8°C and in the worst case up to 4.8°C.

This global warming is leading to a **rise in the sea level**: according to the last IPCC report, from 1901-2010 by an average of 1.7 mm/year, from 1993-2010 by 3.2 mm/year. During the course of the 21st Century, the sea level could rise by between 26 and 82 cm depending on the scenario, whereby the increase of 30-55% of the thermal expansion of the oceans would be responsible for the rest of the glaciers melting.

The **ice at the Poles** has receded at an unexpectedly high rate over the last few years. According to the most recent IPCC report, from 1992-2001 the Greenland ice cap was losing 34 billion tonnes per year, from 2002-2011 215 billion tonnes per year. The Arctic Sea ice coverage reached its negative record in 2012. In 2013, the boundary of the compact pack ice (more than 90% ice coverage) north of the Russian Franz Josef Land and Sewernaja Semlja archipelago receded behind the 88th parallel for the first time since the start of satellite measurements. In the summer of 2014, the melted ice was also above the average for 1981-2010. Snow and ice were also darker than in 2013, which further accelerated the melt, because depending on snow cover, the sea ice reflects sunlight to between 60-90% (albedo effect), while dark snow and ice surfaces absorb it to 90%. This heats the sea water. This not only promotes further melting of the ice, but also the release of the greenhouse gas, methane, from marine sediments. Climate researchers call such effects positive feedback.

In the Antarctic, the ice melt increased five-fold from an annual figure of 30 billion tonnes (1992-2001) to 147 billion tonnes per year in the period from 2002 to 2011. In spring 2014, the West Antarctic ice sheet reached a decisive **tipping point**: The melt led to further destabilisation that accelerates melting and makes the collapse of the ice sheet irreversible, according to the results of several scientific studies.

The **melting of mountain glaciers** has also accelerated in recent years. According to the most recent IPCC report, from 1971 to 2009 on average 226 billion tonnes of ice were lost per year, from 1993 to 2009, the loss increased to 275 billion tonnes per year. In the long term, the lack of glaciers in mountain valleys may lead to water shortages. Unlike Central Europe or North America with their summer precipitation, this would affect the dry regions of Asia, where they are almost exclusively supplied with water from the glacier melt water, which has an impact far into the flatlands, for example in the current catchment area of the Pamir Mountains glacier.

Because a warmer atmosphere absorbs more moisture and contains more energy overall, climate researchers expect an **increase in weather extremes**. In the northern hemisphere, a second effect is also apparent. Climate change is also affecting the Jet-stream along which global air flows in the middle latitudes. Depending on the location, it draws tropical air northward or

Arctic air southward. If the Jet-stream were to "become snagged", this would lead to extreme weather on the ground. An investigation by the Potsdam Institute for Climate Impact Research (PIK) in 2014 showed that this has happened almost twice as often as before since 2000.

The poorest countries in the world are the ones most affected by the consequences of climate change. According to the study prepared the end of 2012 by PIK on behalf of the World Bank, the expected temperature increase of 4°C by 2100 will have a particularly violent impact on the tropical regions. According to this, the expected rise in sea levels around the equator will be 15-20% more severe than elsewhere - which also increases the risks in the event of more intense tropical storms and floods. The future average temperature is above the current level for heat waves; droughts and crop failures will be more frequent and serious."[100]

Climate protection will thus provide protection for the world population in the case of extreme climate change. Let us recap on world political measures to date. "In the 1992 Framework Convention on Climate Change signed at the Earth Conference in Rio de Janeiro (Brazil), 152 states formulated the common goal of "stabilisation of greenhouse gas concentrations in the atmosphere to reach a level at which dangerous anthropogenic interference in the climate system is prevented". The Framework Convention on Climate Change came into force on 21/03/1994 and was ratified until May 2015 by 195 states and the EU. At the World Climate Conference in Cancun (Mexico) at the end of 2010, this target was concretised as binding for the first time: According to this,

---

[100] Der neue Fischer Weltalmanach 2016, p. 694f

warming should be limited to 2°C compared to pre-industrial levels by 2100. According to the latest IPCC report, the cumulative CO_2 emissions must not exceed 2900 billion tonnes in order to achieve this goal. However, 69% of this quantity has already been emitted since the beginning of industrialisation.

According to the IPCC, the **two-degree target** can only be achieved if the CO_2e-concentration in the atmosphere is at 450 ppm in 2100. According to the IPCC, through appropriate climate protection measures, a decrease in consumption by an average of 1.7% in 2030, 3.4% in 2050 and 4.8% in 2100 is to be expected; positive economic effects of climate protection measures, e.g. in the areas of health and the prevention of air pollution, are not included in the calculation. Delaying these measures would lead to higher costs in the long term.

The United Nations Environment Programme (UNEP) Gap Report, which has been published annually since 2011, calculates the CO_2 budget for compliance with the **two-degree target**: According to this, from 2055 to 2070, CO_2 neutrality must be achieved, i.e. the emissions must be compensated for by sinks. From 2080 to 2100, emissions must be reduced to zero."[101]

The World Climate Conference in Lima in 2014 was an important stop on the way to the summit on climate change to be held in Paris in December 2015. "At the UN World Climate Conference in Lima (Peru) from 01-14/12/ 2014, the representatives of 195 countries and the EU continued work on the text of the agreement

---

101   Ibid. p. 695f

to be adopted in Paris in December 2015 and to enter into force in 2020. The legal form the agreement was to take remained open. The main contention was the balance between emission reduction and adaptation. The final document stressed that the latter should play a greater role in the future; this was one of the major concerns of the developing countries.

Representatives of the indigenous organisations of the Brazilian Amazon at the World Climate Conference in Lima

At the same time, it was decided that all states were to submit their planned climate protection contributions as soon as possible. Another key point was the distribution of commitments among the states. The Kyoto Protocol was only compulsory for industrialised countries. These (and the EU) now argue for the distinction to be abandoned and to base the obligation on the economic capacity of the states; this would concern emerging countries such as Brazil or China. Over USD 10 billion were paid into the Green Climate Fund, which is to support developing countries with climate protection and adaptation."[102]

It is now scientifically generally accepted that forests are among the most important carbon sinks. Thus, something has to be done to protect the forests.

"Despite numerous efforts, there is as yet **no global convention on the protection of forests**. The UN Forum on Forests (UNFF), set-up in 2000, agreed on an international agreement on forests in 2007. Although not binding under international law, it was positively acknowledged as the first global agreement on the protection of forests. For the first time, criteria for sustainable forest management were fully and uniformly defined. At the 11th sitting of the forum from 04-15/05/2015 the debates focused on the organisation and future work of the forum. A working group was set up to develop a strategy for 2017-2030 and a four-year work plan for 2017-2020 with support from experts.

Irrespective of the international process, many countries have improved the protection of forests. According to the FAO, about 12% of the forests in protected areas are dedicated to the conservation of biological diversity; the area increased by 1.9% per year from 2000 to 2010.

---

102   Ibid. p. 696 with figure

New incentives to preserve the forests are to be expected from the international climate agreement that is to be negotiated at the end of 2015. Under the name REDD (Reduction of Emissions through Deforestation and Damage), protection of the forests was already formulated as a target at the UN Climate Change Conference in Bali in 2007. Poorer countries should, within the framework of this scheme, receive financial compensation if they protect their rainforests. Later, the programme was extended to include sustainable forest management (REDD+)."[103]

It has been known for a number of years from satellite images that the ice in the Arctic is melting faster in summer than it should be according to scientific calculations based on previous climate warming. Since 2012, robot buoys have been used to take measurements in the Arctic Sea and have in the meanwhile discovered that there is now a positive feedback effect for these finding: "Even in the warmest years, the Arctic in the spring is still under an ice shield. But toward the end of the summer, the water surface area there is twice the size of the Mediterranean Sea. The more extensive this area is, the greater the strike length of the wind, which in turn leads to higher waves: The wind drives the water before it - the further and the longer, the more powerful the hill of water.

If the sea is ice-free, it also absorbs more sunlight. This warms the water up, heats the air and thus intensifies the wind. The waves generated by this can then break up areas of ice the size of

---

103   Ibid. p. 699

Germany within days. This creates more open water, which is favourable to the formation of even bigger waves.

It is unclear exactly how much the individual elements contribute to this feedback loop in the destruction of the ice. One must also ask to what extent the waves delay the freeze again in the autumn. In order to better understand these relationships, a better understanding of the interaction between waves and sea ice is required."[104]

The melting of the Arctic ice, however, does not contribute to the rise in global sea levels. This problem is caused by the melting of the glaciers on Greenland and the high mountains. Added to this is the gradual thinning of the Antarctic ice sheet. The effect is made clear in a headline by the weekly newspaper DIE ZEIT: "**Wir lassen sie nicht untergehen**" (We will not let you go down) with the subtitle "Warum der Pariser Gipfel den Durchbruch im Kampf gegen den Klimawandel bringen könnte" (Why the Paris Conference could bring about the breakthrough in the fight against climate change), an article by Claus Hecking.[105]

---

[104] Harris, Mark: Wellen als arktische Eisbrecher. (Waves as Arctic Icebreakers) In: Spektrum der Wissenschaft, October 2015, p. 72 ff

[105] Hecking, Claus: Wir lassen sie nicht untergehen. (We will not let you go down) In: DIE ZEIT No. 39, of 24/ Ibid. p. 26 with figure (excerpt)

The article is introduced by a striking photo:

The caption reads: "Mädchen vor der Küste der indischen Insel Ghoramara, die vom **Untergang** bedroht ist." (Girl on the coast of the Indian island of Ghoramara, which faces the threat of sinking.)[106]

---

[106] Ibid., with diagrams

The biggest emittents of the world in millions of tons CO_2

Graphic **ZEIT** / Source: IEA 2012

Global CO_2 emission using fossil fuels in billions of tons

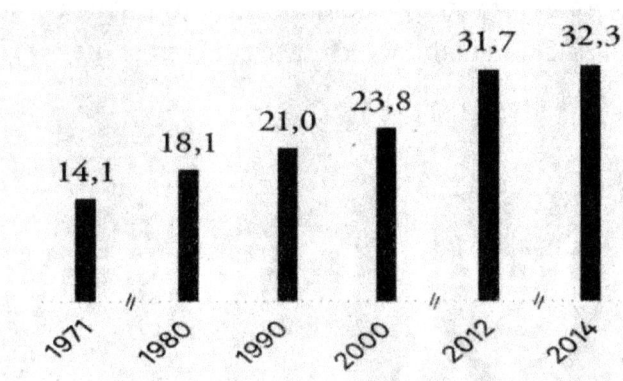

Graphic **ZEIT** / Source: IEA

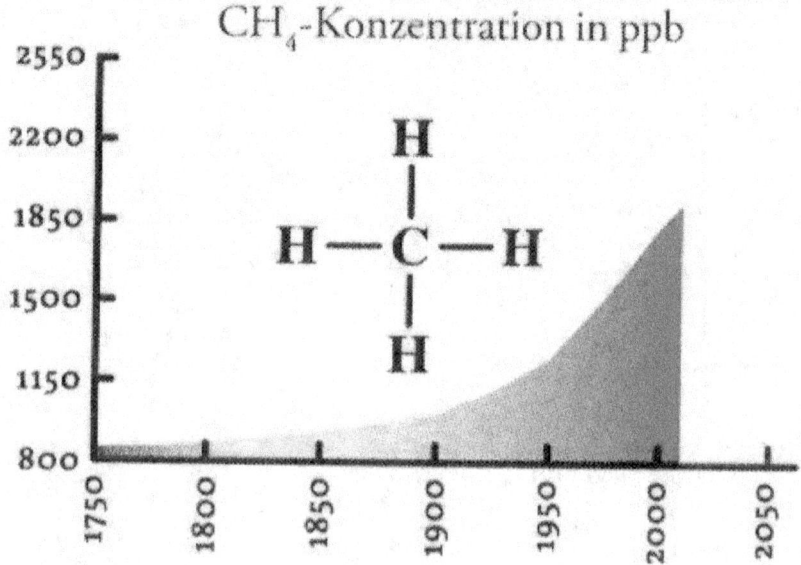

The Society of German Chemists (GDCh) published a 24-page special contribution to the topic of "The Human Planet" in the magazine "Spektrum der Wissenschaft" in October 2015.[107] Among other things, the emissions of climate-damaging gases since 1750 are illustrated: This figure shows the increase in the concentration of methane in the atmosphere up to about 2014. The following figure shows the increase in the carbon dioxide concentration during the same time period:[108]

---

107 GDCh (Ed.) Frankfurt/Main 2015. In: Spektrum der Wissenschaft, October 2015, from p. 86
108 Ibid. p. 5 with figures

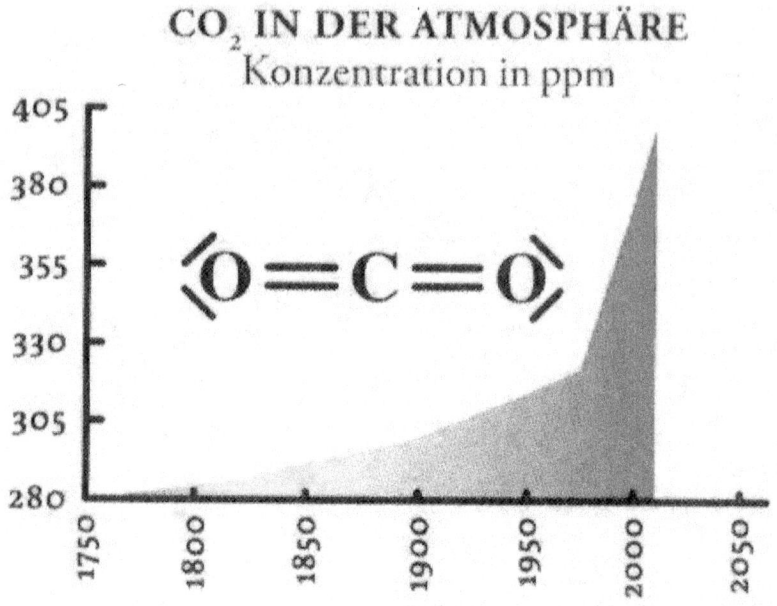

The article **"DIE ENERGIEREVOLUTION"** (The Energy Revolution) with the subtitle "Um die katastrophalen Folgen des Klimawandels abzuwenden, müssen wir bis Ende des Jahrhunderts CO₂-neutral wirtschaften. Experten sind überzeugt: Das geht!"[109] (In order to prevent the catastrophic consequences of climate change, we have to be a CO_2-neutral economy by the end of the century. It can be done!), explains the situation clearly: "The foundations of a low-carbon economy are already in place, the relevant technologies are available. According to the American economist, policy consultant and publicist, Jeremy Rifkin, the energy revolution is also close at hand.

---

109   Ibid. p. 14

What is critical for this, however, is that the political framework conditions in Europe, the USA and the large emerging countries change and investments are provided for this purpose. The insight that this needs to happen obviously exists already.

At the summit of the G7 countries in Elmau Castle, the heads of state and government at least agreed, over the course of the 21st Century, to abandon fossil energies and to achieve "decarbonisation" of the world economy. The IEA surmises that states and companies will have to invest USD 20 billion worldwide in new infrastructure and energy supply systems by 2030. Only eleven trillion would be additionally required to keep climate change within limits using renewable energies.

Therefore, Jeremy Rifkin is also convinced that the "Third Industrial Revolution" will succeed: "In the 21st Century, hundreds of millions of people will generate their own green energy - in their homes, offices, and factories - and share this with others via intelligent distributed electricity networks - internetworks - just as people today create their own information and share it with others via the Internet."[110]

This statement recommends participation in the aforementioned actions in order to prevent multiplication of the costs. If we love our children, grandchildren and great-grandchildren, the well-known maxim "After us, the deluge," repulses us.

Better to participate in shaping the future!

---

110   Ibid. p. 17

6. Paris Climate Conference 2015

In the direction of actively shaping the future, the Climate Conference was held in Paris from 30 November to 12 December. More than 150 heads of state and government came together on the 30/11/2015 to discuss how global warming could be kept within limits.

The group photo shows a greater number of these states leaders[111]

On the Tagesschau TV show on at 17:00 on 30/11, Lorenz Beckhardt, ARD Paris reported as follows: ""We will be deciding on several decades in a few days" according to Hollande. Climate change affects not only life on earth, it also sparks conflicts: "This Climate Conference is about peace", the French head of state emphasised.

Hollande drew hope from the first declarations of intent. More than 180 countries had already issued national declarations of intent in the run-up to the Paris Climate Conference.

---

111 http://www.tagesschau.de/ausland/klima-gipfel-auftakt-103.html

But these intentions must now become actions, the host demanded."[112]

"Chancellor Angela Merkel pushed for a comprehensive and binding climate agreement at the UN Climate Change Conference. Limiting global warming is a "question of the future of humanity", she said in Le Bourget in Paris. For a long time, there has been "the opportunity to reach agreement on our targets for the first time". Transparent methods of measurement must ensure that the promised efforts for climate protection are also verifiable.

Every five years, the commitments made by the individual states should be reviewed. To date, these are not enough to achieve the two-degree target. This should ideally begin before 2020, when the new Agreement is to enter into force. Merkel confirmed the German target of reducing emissions by 40% by 2020 compared to 1990 and by 80-95% by 2050. "Germany will make its contribution", the chancellor promised."[113]

---

112   Ibid.
113   http://www.zeit.de/thema/klimagipfel-2015, with illustration

"Chancellor Angela Merkel demanded that the richer countries realise their financial commitment in particular to the poorer and particularly vulnerable states and make USD 100 billion per year available to them from 2020 for climate protection and management of the impacts of climate change.
Some of the leading economic nations and investors from the private sector have already made initial financial commitments:

- Germany, Norway and the UK want to increase their **finance for forest protection** to a total of **USD 1 billion per year** by 2020. Among others, Brazil, Colombia and Ethiopia could benefit from this. "Forest protection is an important building block of the Paris Agreement," said Federal Environment Minister Barbara Hendricks (SPD). An initial project was already agreed in Le Bourget: Colombia agreed to gradually limit deforestation and to stop it completely by 2020. For the carbon that remains in the trees, the South American country will receive about USD 5 per tonne. According to UN estimates, protection of the forests could achieve approximately one third of the necessary global reduction in greenhouse gases.

- Germany, Norway, Sweden and Switzerland together with the World Bank are founding the **Transformative Carbon Asset Facility (TCAF)**, a new initiative to support **developing countries in the fight against climate change,** with **USD 500 million**. The money will be paid to the countries, for example to provide support in the transition to renewable energy sources or in the fields of energy efficiency and waste management. The initiative began in 2016, initially with EUR 250 million from the founding countries. Until the goal of EUR 500 million had been reached, the programme will remain open to further contributors.

- Canada, Denmark, Finland, France, Germany, Ireland, Italy, Sweden, Switzerland, the United Kingdom and the USA jointly give USD 250 million to the **Least Developed Countries Fund (LDCF)**, a support initiative by the Global Environment Facility (GEF) for developing and particularly vulnerable countries suffering the consequences of climate change. 320 adaptation projects from 129 countries have availed of this aid since 2001. To date, the GEF has paid out a total of USD 1.3 billion from its own resources and been able to mobilise a total of 7 billion from other sources.

- The Chinese head of state Xi Jinping announce the establishment of a **USD 20 billion comprehensive fund** to support **developing countries**. Climate-friendly technologies are to be transferred to developing countries. The needs of these countries to reduce poverty and increase their populations' standard of living must be taken into account.

  The target must be a cooperation with mutual benefit, to which each country contributes what they can.

- US President Barack Obama and French President François Hollande started **Mission Innovation** with Microsoft founder Bill Gates. In this initiative, 20 countries have undertaken to double their **investment in the development of clean technologies** in the next five years. The participating countries include Saudi Arabia, India, China, Indonesia and Brazil."[114]

Merkel and Obama want to concentrate on binding targets at the climate summit.[115]

"The USA want to take responsibility at the climate summit. US President Barack Obama stressed that his country accepts joint responsibility for climate change.

Therefore, the United States have done a lot in past years to expand renewable energies. CO_2 emissions in the USA are now at their lowest level for 20 years.

Federal Chancellor Angela Merkel called for reliability in Paris. Both the agreements as well as later examination of the targets set would have to be binding.

---

114   30/11/2015, 16:23, source: ZEIT ONLINE, dpa, Reuters, AFP, sig
115   Tagesschau 17:00, 30/11/2015, Lorenz Beckhardt, ARD Paris

China also sees renewable energies as an important route to be taken. China is the largest power consumer in the world and produces the most greenhouse gases emissions. The impact of energy consumption is currently clear in the form of smog over the Beijing region. From Xi Jinping's perspective, the climate summit should take into account the different levels of development of the participant countries. Every country should have the opportunity to develop its own solutions to the climate problem. He called upon the industrialised countries to take far-reaching steps... Germany, Norway, Sweden and Switzerland have already started a very specific project together with the World Bank. They want to make USD 250 million available for developing countries. With this money, climate-damaging fossil fuels are to be gotten rid of and legal hurdles to renewable energies reduced."[116]

In the weekly newspaper "DIE ZEIT" of 10 December 2015 on page 1, Claus Hecking describes the reasons why the climate summit in Paris could be successful: Many climate summits have failed in their approach. Thus far, the participants were to agree on a target value for global carbon dioxide emissions and other greenhouse gases and transferred the savings to be made to individual states. "The negotiations often ended in dispute. The Paris organisers have reversed the process - and started a kind of collection for the climate. Each participant is to specify voluntarily how much CO_2 they want to save. Each government only contributes what is beneficial to their country: ecologically and economically.

---

116   Ibid.

The climate collection unites. 185 governments have submitted their contributions, some of which are remarkable. Big polluters like China and the USA, as well as countries such as Ethiopia, announced the development of renewable energies on a grand scale. Behind this a little altruism and a lot of commercial calculation."[117]

The climate summit took longer than planned. It only ended on 12/12/2015 at 19:24.[118]

"The Paris Climate Protection Agreement has been overwhelmingly positive"[119]

---

117 Hecking, Claus: Profit für die Welt. (Profit for the World) In: DIE ZEIT No. 50 of 10/12/2015, p. 1
118 Eckert, Werner: Klimaabkommen von Paris. Ein solides Fundament. (Paris Climate Accord. A Solid Foundation) http://www.tagesschau.de/ausland/klimavertrag-einigung-103.html#header
119 Ibid., with figure

On 12/12/2015 at 22:29, the following report was broadcast on the Tagesschau TV show:

"It is done: The participants in the World Climate Conference in Paris have agreed on a new Climate Protection Agreement. The Agreement, referred to as historic, involved almost all the countries of the world for the very first time in the fight against global warming - unlike the 1997 Kyoto Protocol.

At the UN Climate Change Conference, almost 200 states came to an agreement on the fight against climate change. Without opposition, French foreign minister Laurent Fabius, as the conference chairman, confirmed the decision. "I see the hall, the response is positive, I hear no objections", he said, before the agreement was sealed with a strike of the gavel.

The delegates celebrated the agreement with a standing ovation and minutes of applause. "This is our success, the success of all states involved in this process," enthused the Luxembourg EU Presidency. German Chancellor Angela Merkel spoke of a "sign of hope". The German Environment Ministry Barbara Hendricks spoke on ARD of "a historic moment", but said that "Paris is not the end but rather the beginning of a long journey". In an interview with *tagesthemen* she warned that: "We have to do even better."[120]

In addition, the topics of the day provided the following background:

"With the pact that was adopted the evening after tough

---

120   http://www.tagesschau.de/ausland/klimavertrag-einigung-101.html

negotiations, global warming is to be limited to less than two degrees measured in terms of the pre-industrial era. The agreement should ultimately initiate a complete conversion of the global energy supply and a move away from coal and oil in order to cut out the emission of dangerous greenhouse gases.

Since the national emissions targets to date are insufficient for achieving these goals, they are to be reviewed every five years from 2023. According to further supplementary decision adopted, the first informal survey is to be carried out in 2018. In the second half of the century, greenhouse gas emissions neutrality should be achieved."[121]

Federal Environment Minister, Barbara Hendricks, in an interview with Thomas Roth, Tagesthemen 23:15, 12/12/2015

On 17/12/2015, the weekly newspaper "DIE ZEIT" published the article **"Jubelt nicht zu früh" (Don't celebrate too soon)** by Claus Hecking on page 25.[122]

---

121 Ibid., with figure
122 Hecking, Claus: Jubelt nicht zu früh (Don't celebrate too soon). In: DIE ZEIT No. 51 2015, p. 25

The subtitle read: "Damit das Klimaabkommen von Paris wirken kann, müssen Öl und Kohle teurer werden" (In order for the Paris Climate Accord to work, oil and coal must be more expensive)[123]

"One could intuitively assume that the age of fossil fuels is also coming to an end on the financial markets too. And yet, environmental activists are not celebrating the continuing deterioration in oil and coal prices over a number of months. On the contrary: The fall in prices on Monday, writes the financial portal, Brakingviews, "has taken the shine off the climate deal". Because nothing is as dangerous to the global energy turnaround decided in Paris as low prices for coal and oil."[124]

However, the major banks and insurance companies have decided not to continue investing in fossil fuels. "The investment bank, Goldman Sachs, has just announced that it will invest a total of USD 150 billion in low-emission energy technologies by 2025. Other Wall Street banks such as Morgan Stanley and the Dutch bank Ing-Diba want to provide significantly less or even no credit at all to the coal industry - probably out of self-interest too. "If you are invested in the fossil industry and 195 countries say that they want to decarbonise, this means risks for your portfolio," says IIGCC Director Pfeifer."[125]

In order to prevent more oil and coal-fired power plants being built again due to falling oil and coal prices, there must be state price increases for the emission of greenhouse gases such as carbon dioxide.

---

123   Ibid.
124   Ibid.
125   Ibid.

In Paris, the introduction of a global carbon dioxide tax was not a serious topic. "The resistance from fuel exporters such as Saudi Arabia, Russia and Venezuela was too great. But the agreement mentioned the possibility of pricing. And French President François Hollande said that he could imagine that, by 2020, all 20 leading industrial and emerging countries (G 20) would introduce CO_2 price systems."[126]

At this point, I am reminded of the three-volume work "**Das Prinzip Hoffnung**" (The principle of hope) by Ernst Bloch, published in 1959 by Suhrkamp Verlag. In January 1978, during my studies, I acquired this philosophical work and read to my benefit, after the reconstruction work following Second World War also made it possible for us refugees to pass the Abitur (German Higher School Leaving Certificate) and complete a course of study.[127]

The principle of hope accompanied the reunification of European societies to form the European Union and the international community to form the UN.

This principle should continue to be applied and upheld in order to maintain life on earth and a life worth living. Shaping the future using the ideas of the enlightenment is possible and, as we wish it for our children and grandchildren, should be economically and ecologically beneficial to the life of the world community.

---

126 Ibid.
127 Bloch, Ernst: Das Prinzip Hoffnung (The Principle of Hope). Frankfurt am Main 1959, 4th Edition 1977

List of References

Beckhardt, Lorenz, ARD Paris, Tagesschau 17:00 Uhr, 30.11.2015

Bloch, Ernst: Das Prinzip Hoffnung. Frankfurt am Main 1959, 4. Aufl. 1977

BP, The British Petroleum Company Ltd.: BP statistical review of the world oil industry 1976. London 1977

Burchard, Hans-Joachim: Neue Maßstäbe für ein neues Recht. In: Imhoff/Silenius: Energie – politische Macht. 1976. S. 123 – 131

Der Fischer Weltalmanach 1987, Hg.: Hanswilhelm Haefs, Frankfurt am Main 1986

Der Fischer Weltalmanach 1997, Hg,: Dr. Mario von Baratta, Frankfurt am Main 1996

Der Fischer Weltalmanach 2004, Hg,: Dr. Mario von Baratta, Frankfurt am Main 2003

Der Fischer Weltalmanach 2007, Redaktion: Eva Berié und Heide Kobert (verantwortlich), Frankfurt am Main 2006

Der Fischer Weltalmanach 2010, Redaktion: Eva Berié (verantwortlich), Frankfurt am Main 2009

Der neue Fischer Weltalmanach 2012, Redaktion: Eva Berié (verantwortlich), Frankfurt am Main 2011

Der neue Fischer Weltalmanach 2013, Redaktion: Eva Berié

(verantwortlich), Frankfurt am Main 2012

Der neue Fischer Weltalmanach 2014, Redaktion: Eva Berié (verantwortlich), Frankfurt am Main 2013

Der neue Fischer Weltalmanach 2015, Redaktion: Eva Berié (verantwortlich), Frankfurt am Main 2014

Der neue Fischer Weltalmanach 2016, Redaktion: Christin Löchel (verantwortlich), Frankfurt am Main 2015

DESERTEC: http://www.desertec.org/de/organisation/

Deutschlandfunk: http://www.deutschlandfunk.de/

DIE ZEIT: Das Lexikon in 20 Bänden, Hamburg 2005

DIE ZEIT: http://www.zeit.de/thema/klimagipfel-2015

Evers, Ingo: Nach dem Ölschock: Weltwirtschaft im Umbruch. In: Imhoff/Silenius: Energie – politische Macht. 1976. S. 97 – 122

Fernau, Friedrich Wilhelm: Perspektiven der Erdölversorgung. In: Imhoff/Silenius: Energie – politische Macht. 1976. S. 83 – 96

Fischermann, Thomas: Es läuft wie schlecht geschmiert. In: DIE ZEIT, Hamburg, No 2 2015 S. 25

GDCh (Hg.) Frankfurt/Main 2015. In: Spektrum der Wissenschaft, Oktober 2015, nach S. 86

Harris, Mark: Wellen als arktische Eisbrecher. In: Spektrum der Wissenschaft, Oktober 2015, S. 72 ff

Hecking, Claus: Gemeinsam schnell die Welt retten, in: DIE ZEIT Nr. 32 2015, Hamburg 6.8.2015, S. 23

Hecking, Claus: Jubelt nicht zu früh. In: DIE ZEIT Nr. 51 2015, Hamburg 17.12.2015, S. 25

Hecking, Claus: Profit für die Welt. In: DIE ZEIT Nr. 50 vom 10.12.2015, S. 1

Hecking, Claus: Wir lassen sie nicht untergehen. In: DIE ZEIT Nr. 39 vom 24. 09.2015, S. 26

IRENA: https://de.wikipedia.org/wiki/Internationale_Organisation_für_

erneuerbare_Energien

Krüger, Ralf E., Schultze, Christine: Offshore-Branche schöpft wieder Hoffnung, in Rhein-Neckar-Zeitung / Nr. 181 vom 8./9. August 2015, S. 22

Lexikon der Physik, 2000. Spektrum Akademischer Verlag GmbH Heidelberg, Band 5 S. 348f, Band 4 S. 294f, Band 2 S. 97

Lieser, Peter: Zur Genesis der Energiekrise. Der vierte Nahostkrieg, Erdölpolitik und internationale Beziehungen. In: Orient 1975, Nr. 2 (Juni), S. 21 – 56

Luther, Carsten: Elmau-Gipfel Die G7 allein können es nicht richten, in: http://www.zeit.de/politik/deutschland/2015-06/g7-ergebnisse-kommentar

Masdar: http://masdar.ae/ und https://de.wikipedia.org/wiki/Masdar

Meadows, D. u. a.: Die Grenzen des Wachstums, 1972

Münch, Erwin (Hrsg.): Tatsachen über Kernenergie. Essen 1980.

Quellenangabe [5]: Plasma Physics and Controlled Nuclear Fusion Research, Vols. I und II, IAEA-Wien 1979, insbesondere Eubank, H. Et al., PLT Neutral beam heating results, S. 167

Oktoberkrieg und Truppenentflechtung. Siebte Folge aus: Die Memoiren des Anwar el-Sadat. In: Der Spiegel. Hamburg, 08.05.1978, Nr. 32, 19, S. 201 – 221

Plöger, Sven: GUTE AUSSICHTEN FÜR MORGEN, Frankfurt/Main und München, 2. Auflage 2010

Scherer, Katja: Rein ins Rohr, in: DIE ZEIT Nr. 18, Hamburg 2015, S. 31

Springer, Michael: Wird Fracking den Energiehunger stillen? In: Spektrum der Wissenschaft 8/2014 S. 20

Tagesschau:

http://www.tagesschau.de/ausland/klima-gipfel-auftakt-103.html

http://www.tagesschau.de/ausland/klimavertrag-einigung-101.html

US-Außenministerium u. a.: The Global 2000 Report to the President, Washington 1980, Heraus¬gabe der deutschen Übersetzung: Reinhard Kaiser, bei Zweitausendeins, Frankfurt am Main 1980

Winnacker, Karl/Wirtz, Karl: Das unverstandene Wunder, Kernenergie in Deutschland, Düsseldorf – Wien 1975

ZEIT ONLINE, dpa, Reuters, AFP, sig, 30.11.2015, 16:23 Uhr